Ultraviolet (UV) Disinfection for Water Treatment

Ultraviolet (UV) Disinfection for Water Treatment

Second edition

JAMES R. BOLTON
AND
CHRISTINE A. COTTON

American Water Works
Association

Disclaimer
The authors, contributors, editors, and publisher do not assume responsibility for the validity of the content or any consequences of their use. In no event will AWWA be liable for direct, indirect, special, incidental, or consequential damages arising out of the use of information presented in this book. In particular, AWWA will not be responsible for any costs, including, but not limited to, those incurred as a result of lost revenue. In no event shall AWWA's liability exceed the amount paid for the purchase of this book.

Editorial Manager – Book Products: Melissa Valentine
Copy Editor: Suzanne Snyder
Product Manager: Tony Petrites
Production Manager: Gillian Wink
Compositor: Innodata
Cover Design: Michael Labruyere

Library of Congress Cataloging-in-Publication Data

Names: Bolton, James R., 1937- author. | Cotton, Christine A. (Christine Anne), author.
Title: Ultraviolet disinfection handbook / James R. Bolton and Christine A. Cotton.
Description: Second edition | Denver, CO : American Water Works Association, [2022] | Includes bibliographical references and index. |
Summary: "UV light is usually delivered to drinking water as it flows through a UV reactor (Chapter 5 gives more description of the design of UV reactors). The UV equipment (including one or more UV reactors) is typically applied to filtered water and is often installed after the filter effluent piping recombining before the clearwell. It is very important that the operators are able to control the UV output in the reactor so that at least a minimum UV dose is applied at all times"— Provided by publisher.
Identifiers: LCCN 2021061068 (print) | LCCN 2021061069 (ebook) | ISBN 9781647170592 (paperback) | ISBN 9781613005958 (adobe pdf)
Subjects: LCSH: Water—Purification—Ultraviolet treatment—Handbooks, manuals, etc. | Water quality management.
Classification: LCC TD460 .B65 2022 (print) | LCC TD460 (ebook) | DDC 628.1/62—dc23/eng/20220125
LC record available at https://lccn.loc.gov/2021061068
LC ebook record available at https://lccn.loc.gov/2021061069

Printed in the United States of America

American Water Works Association
6666 West Quincy Avenue
Denver, CO 80235-3098
ISBN 978-1-64717-059-2
ISBN, electronic 978-1-61300-595-8

CONTENTS

PREFACE

In the late 1990s, the use of ultraviolet light (UV) to disinfect water was dismissed in North America* because of the (erroneous) perception that UV could not protect against harmful protozoa, such as *Cryptosporidium parvum* and *Giardia lamblia*. That situation changed dramatically in 1998 when it became clear that UV could inactivate protozoa at very low UV doses. The US Environmental Protection Agency recommends the use of UV disinfection in the new LT2 Enhanced Surface Water Treatment Rules and published the UV Disinfection Guidance Manual in 2006. Hundreds of UV disinfection facilities were designed and built over the last 20 years.

This book is designed to provide a practical introduction to the subject of UV disinfection and UV reactors. It is not a monograph; hence, the literature coverage is not comprehensive, although essential references are given.

No background knowledge is assumed except that of general science, such as that given in undergraduate science and engineering degree programs. This handbook should be of value to engineering and scientific consultants, water treatment operators and managers, students, and faculty members in science and engineering programs that deal with water treatment and staff in government regulatory offices.

Careful attention was paid to terminology, units, and definitions. An appendix provides a convenient summary.

The second edition includes updates over the last 15 years of UV disinfection facility operations and research.

* The use of UV to disinfect water (largely groundwater) has been prevalent in Europe for more than 40 years.

ACKNOWLEDGMENTS

The authors appreciatively acknowledge the United States Environmental Protection Agency for allowing the authors to utilize the Ultraviolet Disinfection Guidance Manual's (UVDGM) text, figures, and tables; in many cases, these were used verbatim from the UVDGM.

We are particularly grateful to Karl Linden (University of Colorado at Boulder), Jim Malley (University of New Hampshire), Todd Elliot (Jacobs), and several Los Angeles Department of Water and Power Treatment operators, all of whom provided insightful comments that improved this book significantly. We have incorporated their comments wherever possible. In addition, Isaac Waite and Chip Blatchley provided insights on sleeve fouling.

Christine Cotton would like to dedicate this book to her son (Ethan) and her daughter (Amelia) who inspire her to be a better person, engineer, and mother. She would also like to thank her husband (Herb) and extended family (Maddie, her mother, and Neal) for their love and encouragement. Jim Bolton would like to thank his wife, Ingrid Crowther, for her unwavering support and encouragement during the writing of this book.

CHAPTER 1
Introduction

WHAT IS ULTRAVIOLET DISINFECTION?

Ultraviolet (UV) lamps emit radiation in the 200–400 nm wavelength range (see Chapter 2 for more details on the nature of UV radiation). The 200–300 nm range is often called *germicidal* because this UV radiation is absorbed by the DNA and/or RNA of microorganisms. The absorption of UV photons causes changes in the structure of the DNA and/or RNA, rendering the microorganisms incapable of replicating. Because they cannot multiply, they cannot cause disease, even though technically they still show metabolic activity.* The degree of microorganism inactivation is determined by the UV dose (or fluence)† applied. More details on the mechanism of UV disinfection are given in Chapter 3.

UV radiation is usually delivered to drinking water as it flows through a UV reactor (Chapter 5 gives more description of the design of UV reactors). The UV equipment (including one or more UV reactors) is typically applied to filtered water and is often installed after the filter effluent piping recombining before the clearwell. It is very important that the operators are able to control the UV output in the reactor so that at least a minimum UV dose is applied at all times.

UV disinfection is a physical disinfection process. No chemicals are added, and there is no residual effect once the water leaves the UV reactor.

* It is for this reason that the word *kill* should not be used in regard to UV disinfection.

† The correct term is *fluence*; however, the term *UV dose* is widely used, particularly in North America. Thus, the term *UV dose* is used in most places in this book. Nevertheless, see the discussion in Chapter 2.

THE SCOPE OF THIS BOOK

The focus of this book is ultraviolet disinfection of drinking water. There is some secondary discussion of ultraviolet disinfection of wastewater, air, and surfaces.

HOW DOES UV DISINFECTION COMPARE TO CHEMICAL DISINFECTION?

UV disinfection does not significantly change the water quality and thus does not form regulated disinfection byproducts (DBPs). Also, recent research (see following section) has demonstrated that UV radiation is effective for almost all bacteria, viruses, and protozoa. In contrast to chlorine-based chemical disinfectants, there is no residual disinfection capacity with UV disinfection. Thus, chlorine or chloramines are often used following UV disinfection to provide a disinfectant residual in the distribution system, when required.

Recently, UV disinfection has received a lot of attention because of the SARS-CoV2 virus crisis. Although UV inactivation of this virus occurs primarily in air or on surfaces, the same mechanism of disruption of DNA or RNA in the virus applies.

HISTORY OF ULTRAVIOLET DISINFECTION

UV radiation has had a long and fascinating history in connection with its effects on microorganisms. The following is a brief chronology:

- 1801: Johann Ritter, a pharmacist in Silesia (now in Poland) discovered UV radiation by demonstrating that silver chloride is decomposed most efficiently by the invisible rays beyond the violet.
- 1842: Becquerel and Draper showed independently that a gelatin solution containing silver iodide darkened when exposed to sunlight (Hockberger 2002). This was the first indication of the spectral extent of UV radiation.
- 1877: Downes and Blunt (1877) observed that test tubes filled with a broth containing bacteria, when exposed to sunlight, eventually became sterile.
- 1878: Downes and Blunt (1878) further observed that, by using color filters, it is the blue and violet end of the spectrum that is responsible for the inactivation of bacteria.
- ~1900: The Danish physician Niels Finsen, considered the founder of modern phototherapy, discovered a UV treatment for *lupus vulgaris*, a form of skin tuberculosis. For this discovery, he received the Nobel Prize in Physiology or Medicine in 1904.
- 1903: Bernard and Morgan found that the most sensitive wavelengths for the inactivation of bacteria are around 250 nm (see Lorch 1987).

- 1904–1905: Hertel showed that UV radiation from arc lamps was far more powerful than sunlight in its effect on microorganisms. He showed that the order of efficacy was UVC > UVB > UVA radiation (see Chapter 2 for a definition of these UV ranges).
- 1904: The glassblower Richard Küch at Heraeus in Germany successfully learned how to blow quartz and was able to fabricate the first quartz enclosed mercury lamp.
- 1910: Henri et al. (1910a, 1910b) described the first commercial UV facility that used UV radiation to disinfect water in Marseilles, France. A diagram of this plant is shown in Figure 1-1. However, because of problems with technical failures, unreliable UV lamps, and instability of the electrical power supply, the system was shut down a short time after it was started.
- 1914: Henri and Moycho (1914) found that 280 nm is the most effective of the wavelengths emitted from mercury arc lamps. (It is now known that the 254 nm emission is more effective.)
- 1920s: Chlorine disinfection was introduced. This technology was much cheaper than UV disinfection, so the latter went out of favor until the 1950s.
- 1938: Westinghouse introduced the first fluorescent gas discharge lamp.
- 1929, 1930: Gates (1929, 1930) was the first to carry out detailed investigations of the action spectrum of *E. coli*; he showed that the optimal wavelength for inactivation was about 260 nm, and that the effectiveness spectrum matched that of the DNA absorbance spectrum.
- 1949: Kelner (1949) was the first to discover *photoreactivation*. He found that bacteria stored for some time after UV exposure were able

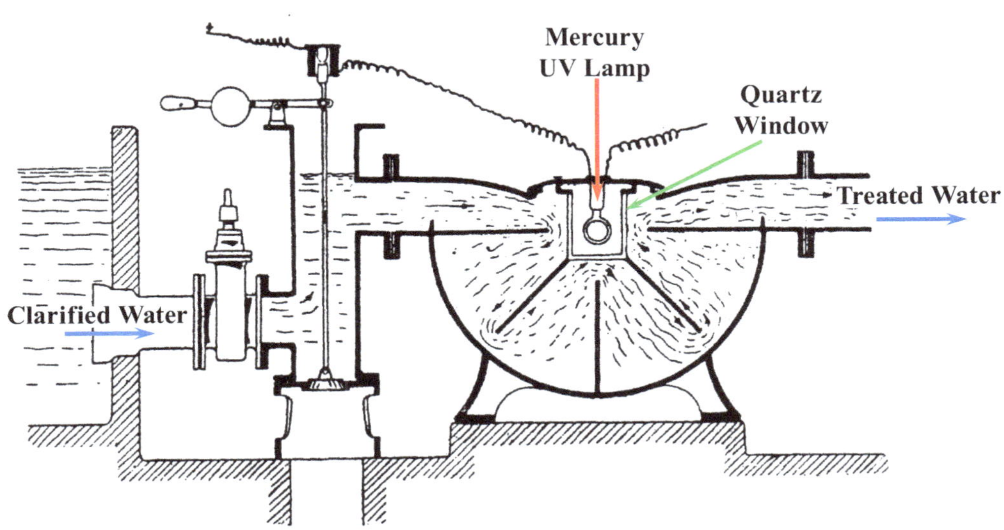

Figure 1-1 UV disinfection plant installed in 1910 in Marseilles, France (Henri et al. 1910a)

to recover; however, the effect was quite variable. On further investigation, he found that exposure of bacteria, previously inactivated by UV radiation, to visible or near UV radiation greatly enhanced the ability of the bacteria to recover.

- 1955: The first modern installations of UV disinfection systems using low-pressure UV lamps in water treatment plants occurred in Switzerland and Austria.
- 1960: Beukers and Berends (1960) irradiated frozen thymine solutions with UV radiation and were able to isolate thymine dimers. This was the basis of the currently accepted concept that UV inactivation is initiated by the formation of thymine dimers from adjacent thymines on a DNA strand.
- 1975: UV disinfection was introduced in Norway as a result of concern with the disinfection byproducts from the use of chlorine disinfection.
- 1978: A full-scale UV system was successfully demonstrated at the N.W. Bergen wastewater treatment plant, Walfwick, N.J. (Scheible and Bassell 1981).
- By 1985: There were over 1,500 UV installations in Europe. Most were for the treatment of groundwater and bank-filtered water.
- Prior to 1998: In North America, until the late 1990s, there was little drinking water application of UV disinfection except in small groundwater systems. This was the result of a perception that UV disinfection was not effective for the treatment of protozoa, such as *Cryptosporidium* or *Giardia*.
- 1998: Bolton et al. (1998) found that UV radiation was very effective against *Cryptosporidium* (and later *Giardia*), contrary to the perception at that time that UV disinfection was not effective. Their paper, presented at the Annual Conference of the American Water Works Association (AWWA), changed this perception.‡ They showed that earlier research on these protozoa was flawed and that if one used an infectivity assay, that is, the ability of these protozoa infect neonatal mice, UV radiation was found to be extremely effective in inactivating these organisms. This indicated that UV disinfection could be used as a broad-spectrum disinfectant capable of inactivating almost all viruses, bacteria, and protozoa. Therefore, since 1999 there has been a remarkable increase in interest in UV disinfection for treating drinking water.
- 2001: Sommer et al. (2002) reported that the number of UV installations in Europe had risen to over 6,000, with most treating groundwater.
- 2003: USEPA issued the first draft Long Term 2 Enhanced Surface Water Treatment Rule (LT2ESWTR), which included UV disinfection as a treatment technique. The USEPA also issued a draft Ultraviolet Disinfection Guidance Manual (UVDGM). This stimulated the installation of large UV disinfection systems throughout North America.

‡ This paper was subsequently published (largely unchanged) in the *Journal AWWA* (Bukhari et al. 1999).

- 2006: USEPA issued the final versions of the LT2ESWTR (USEPA 2006a) and the UVDGM (USEPA 2006b).
- By 2012, there were 389 UV disinfection installations at water treatment plants in North America, 161 in Canada, and 148 in the USA (Wright et al. 2012). The flow rates (and number of systems) for these installations were 0.5–5 MGD (161), 5–50 MGD (100), 50–500 MGD (10), >500 MGD (1).
- In 2018, Validation and Site-Specific Action Spectra Correction Factors for Medium Pressure Ultraviolet Disinfection Systems (Wright et al. 2018) was published, which added to the validation calculations and knowledge base.
- In 2020, USEPA published innovative approaches to validation techniques (USEPA 2020).
- 2020: First full-scale municipal drinking water installation using UV-LEDs at Cumwhinton Water Treatment Works in Carlyle, UK.

UV DISINFECTION OF AIR AND SURFACES

Although this book is aimed primarily at UV disinfection of drinking water, it is worth mentioning that the UV disinfection of air and surfaces is gaining considerable attention driven by the COVID-19 pandemic.

Unfortunately, very little is known about the UV sensitivity of microorganisms in air and even less on surfaces. Usually, it is assumed that the UV sensitivity does not change in moving from water to air or a surface. Also, there are few regulations, so some manufacturers make unsubstantiated claims for the efficacy of their devices.

Nevertheless, there has been promising progress in the UV disinfection of hospital rooms and food and beverages. Readers can review the issues of UV Solutions (https://uvsolutionsmag.com) for articles.

GOVERNMENT REGULATIONS

Several countries have now established regulations for UV disinfection.[§] In 1996, Austria was the first country to introduce UV regulations (initially as guidelines but later enacted as binding regulations [ÖNORM 2001]) by requiring that UV reactors be certified, using biodosimetry testing (involving challenging a UV reactor with a nonpathogenic microorganism) to deliver a UV dose (fluence) of at least 40 mJ cm^{-2} (400 J m^{-2}) using a specific challenge microorganism. The first Austrian UV regulations were restricted to UV reactors containing low-pressure (LP) UV lamps but were later extended to UV reactors containing medium-pressure (MP) UV lamps (ÖNORM 2003).

§ Sommer et al. (2002) review the history of UV regulations, particularly those in Europe.

Germany followed a year later with the Deutscher Verein des Gas- und Wasserfaches (DVGW) Standard (DVGW 2006). This standard is similar to that in Austria, in that it also requires a minimum UV dose of at least 40 mJ cm^{-2} (400 J m^{-2}) using a specific challenge microorganism. The German and Austrian standards have now been combined (DIN 2020) into one joint standard.

The US Environmental Protection Agency (USEPA) has established several rules to address various issues of water treatment. Please note that the LT2ESWTR is the current governing requirements for UV disinfection and that state regulators can enforce more stringent requirements than USEPA. The impacts on UV disinfection are as follows:

- Surface Water Treatment Rule (SWTR) (USEPA 1989).
  These rules were based on the perception that UV disinfection is not effective in the treatment of *Giardia* but that it is effective for virus inactivation. These rules included UV dose tables for virus inactivation.
- A group of regulations collectively called the "Stage 1 Disinfectants and Disinfection Byproducts Rule" (USEPA 1998).
- These rules did not consider UV radiation as a disinfectant technology largely because of the perception that UV disinfection was not effective in the treatment of protozoa (e.g., *Cryptosporidium* or *Giardia*), and UV disinfection does not leave a disinfectant residual.
- LT2ESWTR (USEPA 2006a). This rule lays out specific UV dose tables for *Cryptosporidium*, *Giardia*, and viruses. For UV disinfection applications, the companion Ultraviolet Disinfection Guidance Manual (USEPA 2006b) is a valuable source of UV disinfection information. This rule reversed the previous perception that UV disinfection was not highly effective for protozoa. In addition, this rule reversed the previous perception that it was highly effective for virus inactivation and requires a UV dose of 186 mJ cm^{-2} for 4-log virus inactivation.
- Stage 2 Disinfectants and Disinfection Byproducts Rule (Stage 2 DBPR) (USEPA 2006c).
- Ground Water Rule (GWR) (USEPA 2006d). This rule identifies UV disinfection as a promising technology for virus inactivation, although the UV dose levels for 4-log virus inactivation are much higher than *Cryptosporidium* and *Giardia* published subsequently (see below). There has been research in virus inactivation that found wavelengths below 240 nm are more efficient for virus inactivation (Beck et al. 2014, Beck et al. 2015). UV disinfection of viruses is still evolving as UV sensors are being developed to measure these wavelengths and that be used in validation testing and operation. Another method is to install UV reactors in series or in combination with other treatment techniques to meet the virus inactivation requirements. Recent research studies and new validation methods (USEPA 2020) provide new validation techniques that could be used for validating virus inactivation (see USEPA 2020 and Chapter 4 for details).

- In the LT2ESWTR and Stage 2 DBPR, UV disinfection is classified as one of the microbial options and is a *best available technology* because of the regulatory focus on *Cryptosporidium*. Thus, the USEPA has come full circle from rejecting UV disinfection as a viable treatment technology for protozoa to currently highlighting that it is a *best available* disinfection treatment technology. An important innovation in these rules is the classification of water utilities into *bins* according to the level of *Cryptosporidium* oocysts found in their source waters (Table 1-1). Water utilities in Bin 1, with source waters with relatively low levels are considered *safe* and do not require additional treatment, whereas water utilities in higher bins and *Cryptosporidium* levels have to introduce additional treatment. The various treatment technologies are gathered together into a *toolbox*, and each is assigned a log reduction credit. Therefore, utilities can combine treatment technologies to achieve the level of treatment required for their bin. UV disinfection becomes an attractive and cost-effective treatment option for water utilities with a high level of *Cryptosporidium* in their water, and accordingly these utilities are in Bins 3 or 4. The UV regulations are discussed in more detail in Chapter 4.

Table 1-1 LT2ESWTR bin classification for filtered public water systems (PWSs) (USEPA 2006a)

		Additional Treatment Required			
***Cryptosporidium* Concentration (oocysts/L)**	**Bin Classification**	**Conventional Filtration Treatment (Includes Softening)**	**Direct Filtration**	**Slow Sand or Diatomaceous Earth Filtration**	**Alternative Filtration Technologies**
<0.075	Bin 1	No additional treatment	No additional treatment	No additional treatment	No additional treatment
≥0.075 and <1.0	Bin 2	1 log treatment*	1.5 log treatment*	1 log treatment*	As determined by the state*,‡
≥1.0 and <3.0	Bin 3	2 log treatment†	2.5 log treatment†	2 log treatment†	As determined by the state†,§
≥3.0	Bin 4	2.5 log treatment†	3 log treatment†	2.5 log treatment†	As determined by the state†,¶

* PWSs may use any technology or combination of technologies from the microbial toolbox.

† PWSs must achieve at least 1 log of the required treatment using ozone, chlorine dioxide, UV light, membranes, bag/cartridge filters, or bank filtration.

‡ Total *Cryptosporidium* treatment must be at least 4.0 log.

§ Total *Cryptosporidium* treatment must be at least 5.0 log.

¶ Total *Cryptosporidium* treatment must be at least 5.5 log.

ADVANTAGES AND DISADVANTAGES OF UV DISINFECTION

There are many advantages and disadvantages for UV disinfection, as described in the following section.

Advantages

1. It is a very effective disinfection technology for *Cryptosporidium* and *Giardia*.
2. It does not significantly alter the water quality; that is, no change in Total Organic Carbon (TOC), pH, corrosivity, DBP formation potential, or turbidity.
3. The technology is relatively inexpensive with low capital and operating costs, compared to other disinfection options for protozoa.
4. It is relatively easy to operate (i.e., turn up or turn down) the UV equipment based on changes in water flow, water quality, etc.
5. The UV equipment has a relatively small *footprint* and is usually amenable to retrofit into existing water treatment plants.
6. No chemicals are needed for UV disinfection.
7. Disinfection is very fast. Contact times are in the range of a few seconds.

Disadvantages

1. There is no residual disinfection capacity. Therefore, some level of chlorine or chloramines is usually added to maintain a disinfection residual in the distribution system where disinfectant residual is required by regulations.
2. At present, it is not possible to directly and continuously monitor the UV dose, and operators have to rely on calculated UV doses and secondary measurements (sensor readings, UV transmittance, water flow rates, etc.).
3. Most UV reactors contain mercury lamps, so breakage of UV lamps represents a possible mercury hazard. However, calculations (USEPA 2006b) appear to show that even if the mercury in a lamp were to enter the water completely, the mercury level in the distributed water would still be well below maximum contaminant levels. Lamp breaks do pose risks to operations staff and proper equipment and planning is needed.
4. The electric power supply to the utility could be subject to interruptions, which could cause UV lamps to extinguish for time periods of 1–5 min. This could result in some water not being treated unless the water is diverted to waste. These situations are considered by the USEPA as off-specification events because the UV system would be operating outside of the verified limits of performance.

5 The USEPA UV dose requirements for viruses is very high (186 mJ cm^{-2} for 4-log inactivation), so if virus protection is required, the UV reactors need to be able to be designed, validated, and monitored for virus inactivation. New research may make virus inactivation more viable.

REFERENCES

Beck, S.E.; R.A. Rodriguez; K.G. Linden; T.M. Hargy; T.C. Larason; H.B. Wright. 2014. Wavelength Dependent UV Inactivation and DNA Damage of Adenovirus as measured by Cell Culture Infectivity and Long Range Quantitative PCR. *Environmental Science and Technology*, 48(1):591–598. DOI: 10.1021/es403850b.

Beck, S.E.; H.B. Wright; T.M. Hargy; T.C. Larason; K.G. Linden. 2015. Action Spectra for Validation of Pathogen Disinfection in Medium-Pressure Ultraviolet (UV) Systems. *Water Research*, 70:27–37. DOI.org/10.1016/j.watres.2014.11.028.

Beukers, R.; and W. Berends. 1960. Isolation and Identification of the Irradiation product of Thymine. *Biochimica Biophysica ACTA*, 41:550–551.

Bolton, J.R.; B. Dussert; Z. Bukhari; T.M. Hargy; J.L. Clancy. 1998. Inactivation of *Cryptosporidium parvum* by Medium-Pressure Ultraviolet Light in Finished Drinking Water. In Proc. of the AWWA Annual Conference. Denver, Colo.: AWWA.

Bukhari, Z.; T.M. Hargy; J.R. Bolton; B. Dussert; J.L. Clancy. 1999. Medium-Pressure UV Light for Oocyst Inactivation. *Journal AWWA*, 91(3):86–94.

DIN 19294:1.2020. Devices for the Disinfection of Water Using Ultraviolet Radiation — Part 1: Devices Equipped with UV Low-Pressure Lamps — Requirements and Testing. Joint Austrian/German Standard. DIN (Deutsches Institut für Normung). https://shop.austrian-standards.at/action/en/public/details/667583/OENORM_M_5873-1_2020_01_01.

Downes, A. and T.P. Blunt. 1878. On the Influence of Light Upon Protoplasm. Proceedings of the Royal Society of London, 28:199–213.

Downes, A. and T.P. Blunt. 1877. Researches on the Effect of Light Upon Bacteria and Other Organisms. Proceedings of the Royal Society of London, 26:488–500.

DVGW (Deutscher Verein des Gas- und Wasserfaches). 2006. W 294 – *UV Devices for Disinfection in Drinking Water Supply: Part I: Requirements on Properties, Function and Operation; Part 2: Testing of Properties, Function and Disinfection Effectiveness; Part 3: Ports and Sensors for the Radiometric Monitoring of UV Disinfection Devices—Requirements, Testing and Calibration*. DVGW (German Association for Gas and Water), Bonn, Germany.

Gates, F.L. 1930. A Study of the Bactericidal Action of Ultraviolet Light III: The Absorption of Ultraviolet Light by Bacteria. *Journal of General Physiology*, 14:31–42.

Gates, F.L. 1929. A Study of the Bactericidal Action of Ultraviolet Light II: The Effect of Various Environmental Factors and Conditions. *Journal of General Physiology*, 13:249–260.

Henri, V. and V. Moycho. 1914. Actions des Rayons Ultraviolets Monochromatique dur les Tissues. Mesure de L'Energie de Rayonnement Correspondant au Coup de Soleil. *Comptes rendus hebdomadaires des séances de l'Académie des sciences*, 158:1509–1511.

Henri, V.; A. Helbronner; M. de Recklinghausen. 1910a. Nouvelles Recherches sur la Sterilization de Grandes Quantités d'Eau par les Rayons Ultraviolets. *Comptes rendus hebdomadaires des séances de l'Académie des sciences*, 151:677–683.

Henri, V.; A. Helbronner; M. de Recklinghausen. 1910b. Sterilization de Grandes Quantités d'Eau par les Rayons Ultraviolets. *Comptes rendus hebdomadaires des séances de l'Académie des sciences*, 150:932–934.

Hockberger, P.E. 2002. A History of Ultraviolet Photobiology for Humans, Animals and Microorganisms. *Photochemistry and Photobiology*, 76(6):561–579.

Kelner, A. 1949. Effect of Visible Light on the Recovery of *Streptomyces griseus condidia* From Ultraviolet Irradiation Injury. Proceedings of the National Academy of Sciences of the United States of America, 35(2):73–79.

Lorch, W. 1987. *Handbook of Water Purification*. 2nd ed. Chichester, UK: Ellis Horwood. Malley, J.P., J.P. Shaw, and J.R. Ropp. 1996. *Evaluation of Byproducts Produced by Treatment of Groundwaters With Ultraviolet Irradiation*. Denver, Colo.: AWWA Research Foundation.

ÖNORM (Österreichisches Normungsinstitut). 2003. Plants for the Disinfection of Water Using Ultraviolet Radiation—Requirements and Testing—Part 2: Medium Pressure Mercury Lamp Plants. ÖNORM M 5873-2. Vienna, Austria: ÖNORM.

ÖNORM. 2001. Plants for the Disinfection of Water Using Ultraviolet Radiation—Requirements and Testing—Part 1: Low Pressure Mercury Lamp Plants. ÖNORM M 5873-1. Vienna, Austria: ÖNORM.

Scheible, O.K., and C.D. Bassell. 1981. Ultraviolet Disinfection of a Secondary Wastewater Treatment Plant Effluent. EPA Report No. EPA-600/S2-B1-152, Cincinnati, Ohio: US Environmental Protection Agency.

Sommer, R., A. Cabaj, G. Hirschmann, W. Pribil, and T. Haidler. 2002. UV Disinfection of Drinking Water in Europe: Application and Regulation. In Proc. of the First Asia Regional Conference on Ultraviolet Technology for Water, Wastewater, and Environmental Applications. Scottsdale, Ariz.: International Ultraviolet Association. CDROM.

USEPA (US Environmental Protection Agency). 2020 Wright, H., T. Brooks and M. Heath. 2020. Innovative Approaches for Validation of Ultraviolet Disinfection Reactors for Drinking Water Systems. Report EPA/600/R20/094, US Environmental Protection Agency, Washington, DC. [Online].

USEPA. 2006a. Long Term 2 Enhanced Surface Water Treatment Rule (LT2) [Online].

USEPA. 2006b. *Ultraviolet Disinfection Guidance Manual for the Final Long Term 2 Enhanced Surface Water Treatment Rule* [Online].

USEPA. 2006c. Stage 2 Disinfectants and Disinfection Byproducts Rule (Stage 2 DBP Rule) [Online].

USEPA. 2006d. Ground Water Rule (GWR) [Online].

USEPA. 1989. Surface Water Treatment Rule [Online].

USEPA. 1998. Stage 1 Disinfectants and Disinfection Byproduct Rule [Online].

Wright, H. 2020. Validation and Site-Specific Action Spectra Correction Factors for Medium Pressure Ultraviolet Disinfection Systems, Water Research Foundation. [Online].

Wright, H.; D. Gaithuma; M. Heath; C. Schulz; T. Bagin; A. Cabaj; A. Schmalweiser; M. Schmetzer; J. Finegan-Kelly. 2012. UV Disinfection Knowledge Base. Denver: Water Research Foundation.

Wright, H. and K.G. Linden. 2018. Validation and Site Specific Action Spectra Correction Factors for Medium Pressure UV Disinfection Systems. Denver: Water Research Foundation.

CHAPTER 2

Fundamentals of UV Radiation and Photochemistry*

UV disinfection arises from a photochemical process, in which UV radiation† is absorbed by DNA or RNA and proteins or other biomolecules (see Appendix A for definitions) in the microorganism, causing a disruption in the structure of the nucleic acid. Therefore, an understanding of UV disinfection requires some background in the fundamentals of UV radiation, photochemistry, and photobiology. This chapter is designed to help the reader understand these fundamental concepts, which are discussed in general terms. Terms and symbols will often be introduced without explanation. Specific definitions of terms, symbols, and units are given in Appendix A.

Light is important to almost all life forms. Humans *see* only a very small fraction of the *colors* of light. This book primarily addresses radiation with wavelengths *beyond the violet* end of the rainbow or the *ultraviolet*. There are several references for information on radiation and its measurement; for example, a free handbook is available from International Light (Ryer 1997).

LIGHT AND PHOTONS

Light is a form of electromagnetic energy that extends from radio waves to cosmic rays over at least 16 orders of magnitude in wavelength. Radiation has both particle and wave properties. It is transmitted in discrete packets of energy (photons) and yet has a frequency and wavelength. The connection between these two properties is embodied in the Planck Law of Radiation

* Some of this chapter has been adapted from Bolton (2015).

† Physicists prefer to use the term *ultraviolet radiation*, reserving the term *light* for *visible light*. The authors will adhere to this definition, except where the term *light* implies a broader context.

$$u = h\nu = hc/\lambda = hc\bar{\nu} \quad \text{(Eq 2-1a)}$$

$$U = N_A h\nu = hcN_A/\lambda = hcN_A\bar{\nu} \quad \text{(Eq 2-1b)}$$

where u is the *energy* (J) of one photon; ν is the *frequency* (s^{-1}), λ is the *wavelength* (m); $\bar{\nu}$ is the *wavenumber* (m^{-1}); c is the *speed of light* (2.9979×10^8 m s^{-1}) in a vacuum; h is the *Planck constant* (6.6261×10^{-34} J s); N_A is the *Avogadro number* (6.02214×10^{23} mol^{-1}); and U is the *energy per einstein*.‡ The units here have been given in the standard Système Internationale (SI) forms (see Appendix A); however, for applications in UV light and photochemistry, λ is usually given in nanometers (nm) and $\bar{\nu}$ in cm^{-1}, with appropriate numerical factors to make Eq 2-1b come out to J $einstein^{-1}$.

The absorption of photons by a molecule raises that molecule to an excited state from which a chemical reaction may occur. (This process is called *photochemistry*. See the following section.)

UV RADIATION SPECTRAL RANGES

The usual wavelength range in photochemistry is 100–1,000 nm. Light photons with wavelengths longer than 1,000 nm have a photon energy too small to cause chemical change when absorbed. Photons with wavelengths shorter than 100 nm have so much energy that molecules are ionized, producing electrons and reactive radicals (e.g., ·OH), which cause molecular disruptions characteristic of the effects of radiation from radioactive sources. The total photochemical wavelength range is divided into bands with specific names as shown in Figure 2-1.

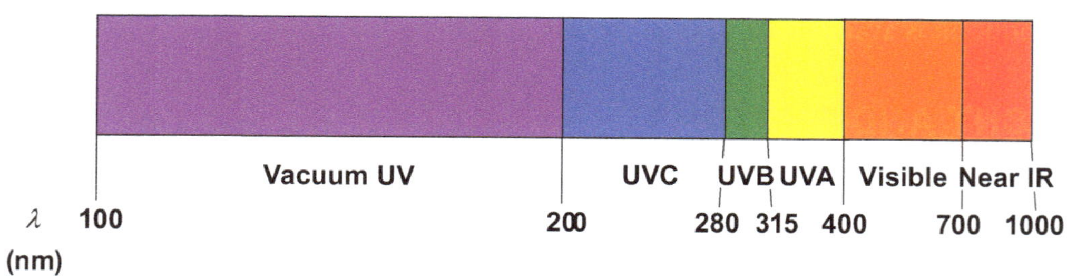

Figure 2-1 Spectral ranges of UV, visible, and IR radiation. The ultraviolet range is 100–400 nm§

‡ An *einstein* is one mole (6.0221408×10^{23}) of photons.

§ UVC is often defined as 100–280 nm; however, we choose to split the range because of the sharp change of properties in the vacuum ultraviolet (VUV).

Little photochemistry occurs in the *near infrared* (700–1,000 nm), except for some photosynthetic bacteria, which are capable of storing solar energy at wavelengths out to 980 nm. The *visible* range (400–700 nm) is completely active for photosynthesis in green plants and algae. Also, many dyes undergo photochemical transformations in the visible range or sensitize reactions in other molecules. In addition, some research showed wavelengths between 300 and 500 nm can inhibit nitrification in marine waters (Guerrero and Jones 1996 and Olson 1981), and some water utilities have evaluated the use of UV radiation to reduce nitrification in chloraminated distribution systems (Wilczak and Lai 2006 and White and Adams 2008).

Most studies in photochemistry involve the *ultraviolet* ranges (100–400 nm). This region is divided into four subranges connected with the human skin's sensitivity to UV radiation. The UVA range (315–400 nm) causes changes in the skin that lead to suntanning. The UVB range (280–315 nm) can cause sunburning and eventually can induce skin cancer. The UVC range (200–280 nm) is absorbed entirely in the atmosphere, so UVC radiation from the sun does not impact humans. However, UVC from UV lamps or other devices is absorbed by RNA, DNA, and proteins and can lead to cell mutations, cancer, or cell death. In regard to human exposure, UVC must be avoided with appropriate safety measures. UVC is absorbed strongly in the surface layer of skin cells, which are sloughed off almost daily. Therefore, moderate exposures can be tolerated. However, UVC exposure to the eyes is more dangerous because it can cause the formation of cataracts (Wieringa 2006).

There is some evidence that UV radiation in the Far-UV region (200–225 nm) is not very harmful concerning skin and eye exposure (IUVA 2021). Some companies are marketing equipment for UV disinfection of air and surfaces where the UV also exposes humans.

The UVC range is sometimes called the *germicidal range* because it is very effective for inactivating protozoa, bacteria, and viruses. The *vacuum ultraviolet* (VUV) range is absorbed by almost all substances (including liquid water and oxygen in air). Therefore, it can only be transmitted in a vacuum.

EMISSION, TRANSMISSION, AND ABSORPTION OF RADIATION

Various aspects of radiation are discussed in the context of *emission* from a UV lamp, *transmission* through a medium (air, water, etc.), and *absorption* by a target molecule.

Light Emission

The radiation emission from a UV lamp depends on a number of factors, such as the electrical power input, the radiant power efficiency, the dimensions of the lamp, the amount of mercury in the lamp, and the characteristics of a quartz sleeve (if present). The *brightness* of the lamp is described by the *emittance* (W m^{-2}), which is the radiant power emitted per unit area from a small area of the lamp.

Light Transmission and Absorption

Light is transmitted at a speed of 2.9979×10^8 m s^{-1} in a vacuum but slows down when entering any finite medium. The ratio of the speed of light in a vacuum to that in a given medium is called the *refractive index*, the value of which in a vacuum is 1.000 (the refractive index of air is very close to unity [1.0003]). When light radiation passes from one medium of refractive index n_1 to another with refractive index n_2 ($n_2 > n_1$), two effects occur:

1. Some of the radiation is reflected back from the interface¶ such that the angle of reflection (θ_r) is equal to the angle of incidence (θ_i). For radiation directly incident on a surface, the reflection coefficient is given by

$$R = \frac{(n_2 - n_1)^2}{(n_2 + n_2)^2} \quad \text{(Eq 2-2)}$$

If the two media are air and pure water, $R = 0.025$ at 254 nm.

2. Light entering the new medium is *refracted* so that its direction is changed; that is, the angle to the normal of the surface (θ_2) is less than the angle of incidence (θ_1). The magnitude of these two effects increases as $n_2 - n_1$ becomes larger. These effects are illustrated in Figure 2-2.** This phenomenon can have a beneficial effect by focusing the UV radiation as it passes through an air/quartz/water interface at the quartz sleeve in a UV reactor.

If there are absorbing substances in the medium, the radiation is attenuated as it passes through according to the Beer-Lambert law.

$$T = 10^{-A} = \frac{E^t}{E^0} \quad \text{(Eq 2-3a)}$$

or

$$A = \log\left(\frac{E^0}{E^t}\right) = -\log(T) = al \quad \text{(Eq 2-3b)}$$

E^t and E^0 are the transmitted and incident irradiances†† as a beam passes through a medium over a path length l; A is the *absorbance* (unitless) and the

¶ This is the familiar phenomenon of seeing the *reflection* of a mountain on the surface of a lake. The amount of reflection depends on the angle of incidence, with more light being reflected for more oblique angles.

** For a quantitative discussion of reflection and refraction, see Bolton (2015).

†† The UVDGM (USEPA 2006) uses UV intensity (symbol I) in place of irradiance.

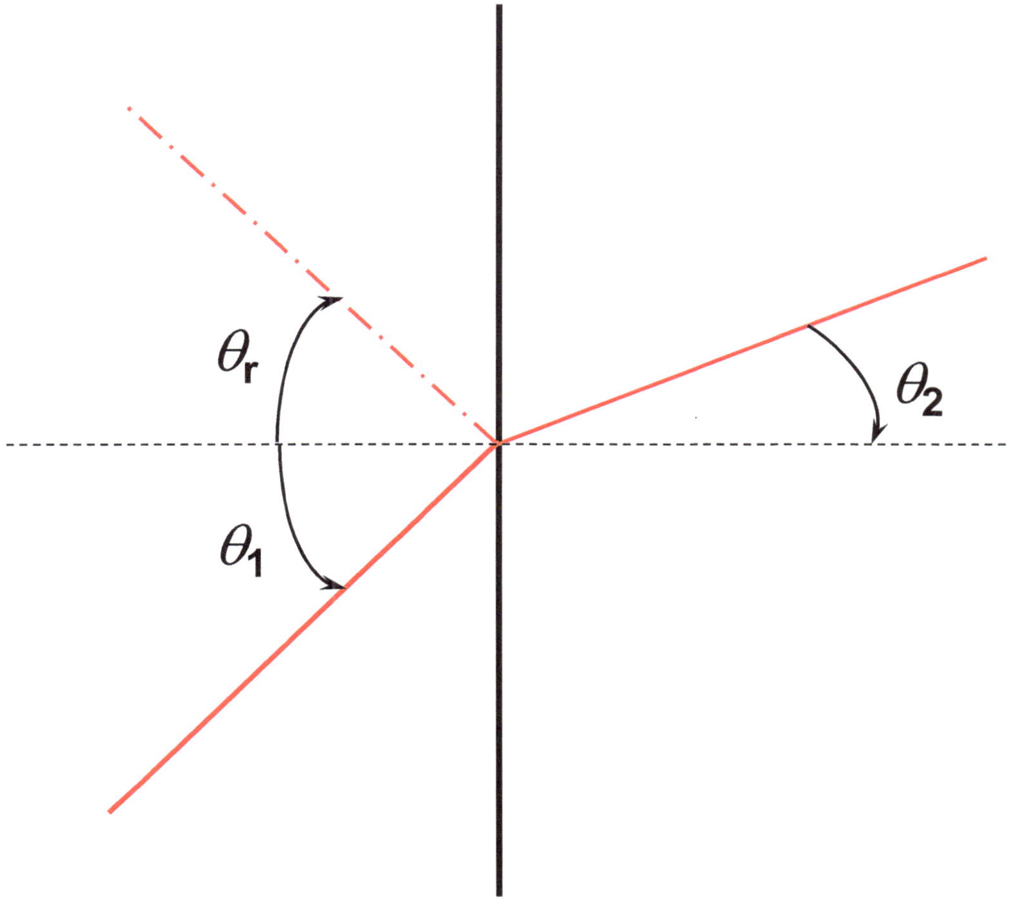

Figure 2-2 Reflection and refraction as a light beam passes from medium 1 with refractive index n_1 to medium 2 with refractive index n_2. The dotted line represents the reflected light. θ_1 is the angle of incidence, θ_r (= θ_1) is the angle of reflection and θ_2 (< θ_1) is the angle of refraction.

fraction T is the *transmittance* (unitless).[‡‡] A is directly proportional to the path length l, that is, $A = al$, where a is the *absorption coefficient* (cm^{-1}). It is important to understand that Eqs. 2-3a and 2-3b apply only for a beam of light with a narrow range of wavelengths. In fact, a plot of A versus wavelength is called the *absorption* or *absorbance spectrum*, and a plot of T versus wavelength is called the *transmittance spectrum*.

Eqs 2-3a and 2-3B assume that no scattering occurs in the beam. If scattering by particles is present, the situation is more complex because some of the radiation is scattered at oblique angles and some may be absorbed by the particles.

The absorbance is related to the concentrations of absorbing components by

‡‡ Often the transmittance is expressed as the *percent transmittance* (%T), which usually implies a 1-cm path length.

$$A = \sum_i \varepsilon_i c_i l \qquad \text{(Eq 2-4)}$$

where ε_i is the *molar absorption coefficient* (M^{-1} cm^{-1}) for component i at wavelength λ; c_i is the *concentration* (M) of component i in the solution; and l is the path length (cm). Remember that ε and hence A are wavelength dependent. Eq 2-4 is employed in measurements of absorbance that are often used to determine concentrations.

Often the transmittance of UV radiation in the medium (e.g., drinking water) is described by the UV transmittance (UVT), which is defined as the percent transmittance in the medium when the path length is 1 cm and the wavelength is 254 nm, that is,

$$\text{UVT} = 100\ T = 100 \times 10^{-A(254)} \qquad \text{(Eq 2-5)}$$

where $A(254)$ is the absorbance[§§] at 254 nm in a 1.0 cm path length. The UVT[¶¶] of a drinking water is the most important water quality factor as regards the disinfection efficiency of a UV reactor.

Absorbance and transmittance are measured with a spectrophotometer, which measures the UV irradiance as it passes through a quartz cell containing the solution of interest (e.g., a drinking water sample). Because a small fraction of the UV beam is reflected from the quartz surfaces, a *blank* measurement must be made with only pure solvent (e.g., distilled water) in the quartz cell. Thus, in practice the ratio E^t/E^0 in Eq 2-3a is the ratio of the detector level for the cell with the solution of interest to that for the cell containing pure solvent. The *transmittance* is displayed (or plotted) as this ratio itself. In the case of *absorbance*, the spectrophotometer performs the negative logarithm of this ratio. Because many UV reactors use low-pressure lamps, which primarily emit at 254 nm, the measurement of the absorbance at 254 nm [$A(254)$] has become a standard water quality parameter, even when medium-pressure UV lamps are used. There exists a standard method for $A(254)$ (APHA et al. 1998). This method is typically used for SUVA measurement and includes lab filtration of the sample before analysis, which is *not* recommended when assessing $A(254)$ for UV disinfection purposes.

Most filtered drinking waters have UVT values between 75–95 percent or more. This means that E^t is almost the same as E^0. Significant errors can occur if one uses a 1.0-cm path length cell at high UVT values above 95 percent. To minimize these errors for highly transmitting waters, it is advisable to use a 5.0 cm [$A(254) < 0.050$] or 10.0 cm [$A(254) < 0.025$] cell. $A(254)$ is the measured absorbance A divided by the path length l.

§§ This is really an *absorption coefficient* because the path length is defined as 1 cm.

¶¶ Here the UVT is used, as defined by USEPA (2006), namely, the percent transmittance when the path length is 1 cm.

Light Impinging on a Target

When the light arrives at a final target (e.g., a surface or a microorganism), one must consider the light's intensity. Unfortunately, the word intensity can mean many things, and so it is best to use this term in a qualitative sense, such as to observe that sunlight is more intense than moonlight. The UVDGM (USEPA 2006) has chosen to use the term intensity or UV intensity to mean irradiance or fluence rate.

Irradiance

When quantitative analyses are made, the geometry of the target becomes important. In the case of a flat surface target, the term *irradiance* ($W\ m^{-2}$)*** (symbol E) is used. Irradiance is defined as

> *The UV power incident on an infinitesimally small area dA of a flat surface from all upward directions divided by dA.*

Thus, irradiance is appropriate when the UV shines on a large, nearly flat surface, such as in the UV curing of inks and coatings or in collimated beam experiments (see Chapter 3).

Fluence Rate

In the case of a tiny spherical or almost spherical target, such as a microorganism, the term fluence rate ($W\ m^{-2}$) (symbol E_o) is used. Fluence rate is defined as

> *The UV power incident on an infinitesimally small sphere of cross-sectional area dA divided by dA.*

It should be noted that irradiance and fluence rate have the same units ($W\ m^{-2}$ or $mW\ cm^{-2}$); the only difference is the geometry of the target and the incident UV radiation.

In many situations (e.g., in a collimated beam), fluence rate and irradiance are almost the same. Because the term irradiance is commonly used in the water treatment industry, in this book, that term will usually be used, even though the term fluence rate might be more appropriate. However, in modeling studies, where the details of the radiation distribution are important, the term fluence rate should be used.

Fluence or UV Dose

When the fluence rate is integrated over time, one obtains the fluence or commonly called the UV dose ($J\ m^{-2}$) (symbol F),††† that is,

---

***In North America and some other countries, irradiance is measured as $mW\ cm^{-2}$. Note that $1\ mW\ cm^{-2} = 10\ W\ m^{-2}$.

††† Again, in North America and some other countries, fluence or UV dose is measured as $mJ\ cm^{-2}$. Note that $1\ mJ\ cm^{-2} = 10\ J\ m^{-2}$. The UVDGM (USEPA 2006) uses the term *UV dose* (symbol D) to mean fluence.

$$F = \int E_o dt \tag{Eq 2-6a}$$

If the fluence rate is constant in time

$$F = E_o t \tag{Eq 2-6b}$$

The fluence (UV dose) is a very important factor in UV disinfection because the higher the fluence, the higher is the pathogen reduction.

Fraction of UV Radiation Absorbed

Because photochemistry is initiated only by the photons that are absorbed by the medium, it is important to determine the fraction of radiation absorbed from an incident beam.

It should be noted that all of the equations related to the Beer-Lambert law apply only for monochromatic radiation or for a narrow band of wavelengths where the absorption coefficient does not change significantly with wavelength. For monochromatic or narrow-band radiation in the absence of reflection, the fraction of radiation absorbed $f(\lambda)$ in the medium is given by

$$f(\lambda) = 1 - T(\lambda) = 1 - 10^{-A(\lambda)} \tag{Eq 2-7a}$$

In the case of a single microorganism, where the path length is about 1 μm or less, the absolute absorbance is very small (0.002 or less). Therefore, the fraction of UV radiation absorbed is only 0.1 to 0.2 percent or less. In this case, Eq 2-7a may be expanded as a Taylor series to obtain

$$f(\lambda) \cong \ln(10) A(\lambda) \tag{Eq 2-7b}$$

Under these conditions, the fraction of UV radiation absorbed is directly proportional to the absorbance. This has an important consequence, in that under these conditions, all absorbers in the cell receive the same fluence rate so shadowing by one absorber over another is not possible.

When the radiation source spans over a broad range of wavelengths, one can determine the total fraction of radiation absorbed f_{total} by dividing the spectral band of interest (e.g., 200–300 nm) into several narrow bands (e.g., 1 nm in width). f_{total} is then determined as a weighted average from Eq 2-8

$$f_{\text{total}} = \frac{\sum_{i=0}^{n} w_i \left(1 - 10^{-A(\lambda)}\right)}{\sum_{i=0}^{n} w_i} \tag{Eq 2-8}$$

where $A_i(\lambda)$ is the absorbance in band i and w_i is a weighting factor proportional to the relative emittance of the radiation source in band i. Eq 2-8 is important in systems that use medium pressure UV lamps.

Table 2-1 Comparison of terms, units, and symbols in use in various jurisdictions

Term	Jurisdiction		
	IUPAC	USEPA* and North America	Europe and Other Parts of the World
Intensity	Irradiance† E (W m^{-2})	Intensity‡ I (mW cm^{-2})§	Irradiance E (W m^{-2})
Intensity	Fluence rate† E_o (W m^{-2})	Not defined	Fluence rate E_o (W m^{-2}) or Irradiance E (W m^{-2})
UV dose	Fluence† F (J m^{-2})	UV dose D (mJ cm^{-2})§	Fluence or UV dose H_o (J m^{-2})

* As specified in the Ultraviolet Disinfection Guidance Manual (UVDGM) (USEPA, 2006).

† As defined in this chapter.

‡ In the UVDGM, sometimes intensity means irradiance, but at other times, it appears to mean fluence rate. Unfortunately, the UVDGM does not clearly distinguish the concepts of 'irradiance' and 'fluence rate' as defined in this chapter.

§ 1 mW cm^{-2} = 10 W m^{-2}; 1 mJ cm^{-2} = 10 J m^{-2}.

Terms, Units, and Symbols

Unfortunately, when it comes to terms, units, and symbols for UV radiation parameters, there exists considerable confusion around the world. The International Union of Pure and Applied Chemistry (IUPAC) has put forward a rational set of terms, definitions, units, and symbols (Braslavsky 2007); however, the engineering and science community has been slow to adopt these recommendations. In particular, the UVDGM (USEPA 2006) has gone on its own to present an independent set. Table 2-1 compares the various common uses of terms, units, and symbols so that the reader can make appropriate conversions. There has been debate on the definitions of UV intensity and UV irradiance, and it is recommended that the reader use the terms described in this book. Also, the UVDGM does not distinguish the concepts of irradiance and fluence rate, as discussed in this chapter.

Furthermore, the UVDGM uses UV dose rather than fluence because of the drinking water community's familiarity with the concept of dose. However, dose is almost always used in other contexts to mean total absorption of radiant energy. Other scientific communities (Braslavsky 2007) recommended the term fluence, rather than UV dose. Nevertheless, because UV dose is widely used, particularly in North America, UV dose is used in this book, but sometimes fluence is added in brackets.

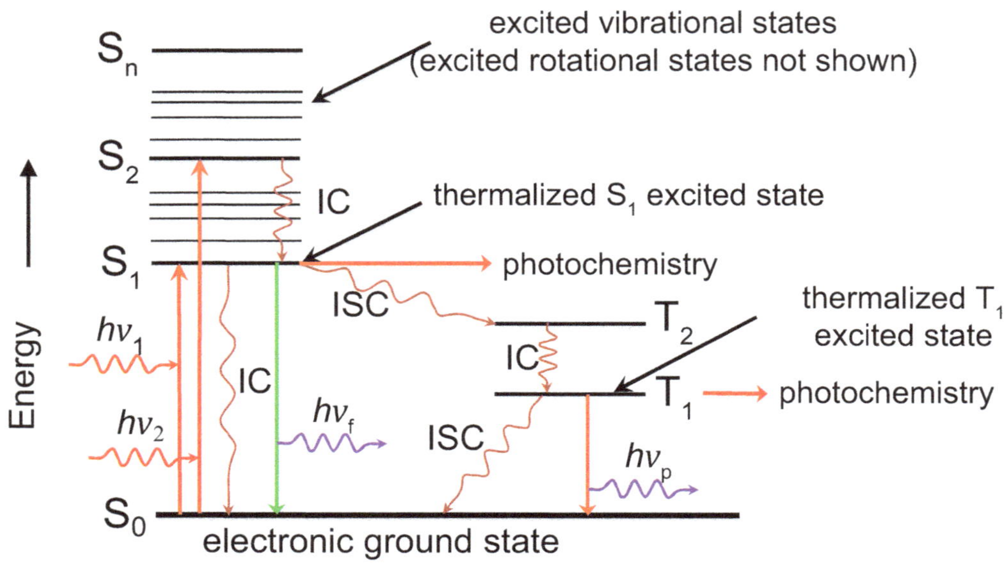

Figure 2-3 Energy levels of a molecule in its ground state and several excited states showing the fundamental processes (adapted from Bolton et al. 2015)

In this book, both the IUPAC and the UVDGM terms, units, and symbols are used wherever possible. Appendix A gives a more complete set of terms, units, and symbols.

WHAT HAPPENS WHEN UV RADIATION REACHES ITS INTENDED TARGET

When UV radiation reaches its intended target, part of it will be absorbed by molecules in the target (e.g., by DNA in a microorganism). The absorbed photon energy causes the molecule to rise in energy from its initial singlet[‡‡‡] ground state S_0 to one or more excited states (S_1, S_2, …) (see Figure 2-3). At this point, several processes can occur:

1. Within a few femtoseconds, the molecule in excited singlet states quickly loses energy to form a *thermalized* excited state S_1.
2. The molecule in the thermalized S_1 state can return to its ground state S_0 with emission of a photon of energy $h\nu_f$—this is called *fluorescence*.
3. The excited state can convert to another (usually longer lived) excited triplet state (T_1, T_2, …) from which photon emission can occur—this is called *intersystem crossing* followed by emission of a photon of energy $h\nu_p$—this is called *phosphorescence*.

‡‡‡ A *singlet state* is a molecule with all electron spins paired, whereas a *triplet state* has two unpaired electrons.

4. The molecule in either the S_1 or T_1 excited states can return to the ground state with the excess energy being lost as heat into the medium—this is called *internal conversion* (IC).
5. The molecule can undergo a chemical transformation from either the S_1 or T_1 excited states (rearrangement or fission)—this is the essence of *photochemistry*.

Photochemistry from an excited state is the principal initial step in UV disinfection. For example, thymine in DNA can be excited by photon absorption, which leads to the formation of a thymine dimer (see Figure 3-4).

The fraction of excited states that result in photochemistry (the rate of step 5 over the sum of the rates of steps 2–5) is called the *quantum yield* (symbol Φ) defined as

$$\Phi = \frac{\text{mole of product formed or reactant removed}}{\text{einsteins of photons absorbed}} \qquad \text{(Eq 2-9)}$$

Note that the denominator is photons absorbed, not incident. Recall that an einstein is one mole (6.022×10^{23}) of photons.

The importance of the *quantum yield* is that it represents the efficiency of the photochemical process. A quantum yield of 1.0 means that all absorbed photons go towards the photochemical conversion.

LAWS OF PHOTOCHEMISTRY[§§§]

The possible photochemical reaction by molecules is governed by several laws of photochemistry.

First Law of Photochemistry

The first law of photochemistry, also known as the Grotthuss-Draper law,[¶¶¶] states that

> *Light must be absorbed by a chemical substance in order for a photochemical reaction to take place.*

This means that molecules that do not absorb radiation of a particular wavelength will not undergo a photochemical reaction when exposed to radiation at that wavelength. For example, when one puts on sunscreen on the skin, no sunburn occurs because the damaging UV radiation is absorbed by the components in the sunscreen. Very little UV penetrates the skin, so no sunburn occurs.

§§§ An excellent textbook on photochemistry is that by Wayne (1988).

¶¶¶ In 1841, John William Draper developed this law. It came to be known as the Grotthuss-Draper law (his name is linked with Grotthuss, who apparently promulgated the same idea in 1817).

Second Law of Photochemistry

For each photon of radiation absorbed by a chemical system, only one molecule is activated for a photochemical reaction.

This law means that each photon of radiation can (at most) cause the photochemical reaction of only one radiation-absorbing molecule. A related law states that the total amount of photoreaction that occurs is directly proportional to the product of the absorbed photon flow and the time of illumination. In other words, more radiation produces more photoproduct.**** This is the reason that UV dose or fluence is so important in the UV disinfection of microorganisms. The UV dose or fluence is directly proportional to the total number of UV photons absorbed by a given microorganism.

Third Law of Photochemistry††††

The energy of an absorbed photon must be equal to or greater than the weakest bond in the molecule.

A photochemical reaction usually involves the breaking of one or more bonds in the molecule and possible reformation of new bonds. Thus, the law of conservation of energy requires that the energy of the absorbed photon must be at least as large as that of the weakest bond broken. For example, when considering the photolysis of nitric oxide (NO_2), the brown gas that is seen so often in the smog over many cities in the summer, absorption of radiation can cause the N–O bond to break in NO_2. However, this bond has a high energy so that even though NO_2 absorbs radiation out to 700 nm, only UV radiation with wavelengths less than about 390 nm can lead to the photochemical breakup of NO_2. radiation of longer wavelengths is absorbed, but the excited state simply returns to the ground state, releasing the energy as heat.

A corollary of the first and third laws is that for a molecule with a very weak bond, a photochemical reaction is only possible if there is significant absorption, even though by the third law, breaking of a bond is theoretically possible. For example, the O–O bond in H_2O_2 is quite weak, such that (theoretically) photochemical fission into two ·OH radicals could occur at wavelengths as long as about 560 nm. However, H_2O_2 does not absorb significantly for wavelengths >300 nm. So, in practice, the photolysis of H_2O_2 only occurs below 300 nm. This is the basis of the UV/H_2O_2 advanced oxidation process (see next section), which is becoming of considerable importance in the treatment of trace contaminants (e.g., herbicides and pesticides) in drinking water sources.

**** This law breaks down for very high photon flows, such as that from a very powerful laser. Under such conditions, the concentration of excited states is high enough so that excited states themselves can absorb a photon, and so multiphoton absorption is possible. This condition is almost never encountered in UV disinfection reactors.

†††† This law applies only for photochemical reactions that involve splitting of the molecule into two fragments. Photo-rearrangement reactions are not subject to this restriction.

ADVANCED OXIDATION TECHNOLOGIES

Advanced Oxidation Technologies (AOTs) are those that utilize powerful oxidizing intermediates (e.g., the hydroxyl radical ·OH) to oxidize primarily organic pollutants from contaminated air and water. The term advanced is used because the chemical oxidation reactions involved are essentially the same (except billions of times faster) as those that take place very slowly if these organic pollutants are dispersed into the environment. Some of the commercially viable AOTs use UV and visible radiation to generate ·OH radicals. Recently, the UV/chlorine process has become another AOT (De Latt and Stefan 2018). Other non-UV based AOTs also exist, such as the O_3/H_2O_2 process. Two other reviews may be consulted for further coverage (Bolton and Cater 1994, Legrini et al. 1993). Finally, books by Oppenländer (2003), Parsons (2004), Collins and Bolton (2016), and Stefan (2018) provide excellent overviews of AOTs.

Treatment with AOTs leads not only to the destruction of the target pollutants but also, given sufficient treatment time and energy, to the complete mineralization (i.e., the only products are CO_2, H_2O, and mineral acids for any Cl, N, S, etc. present in the pollutants) of the pollutants and their byproducts. AOTs have proven to be very effective in treating a wide variety of organic contaminants, such as trichloroethylene, perchloroethylene, 1,4-dioxane, methyl-tert-butyl ether (MTBE), acetone, phenols, N-nitrosodimethylamine (NDMA), benzene, toluene, ethyl benzene, and xylenes (BTEX) found in waters contaminated with gasoline, and many other contaminants in groundwaters and industrial effluents. Also because fluences used for AOT applications are much higher than those used for UV disinfection, almost complete UV disinfection is achieved in any AOT application.

One of the most popular and effective AOTs is the UV/H_2O_2 process, which is based on the direct photolysis of added hydrogen peroxide:

$$H_2O_2 + h\nu \rightarrow 2{\cdot}OH \qquad \text{(Eq. 2-10)}$$

The quantum yield for generation of •OH radicals is 1.1, and most organic pollutants can be degraded rapidly. Because the molar absorption coefficients of H_2O_2 are low, sufficient H_2O_2 must be added (usually >5 mg/L) so that a significant fraction of the UV between 200 and 300 nm is absorbed. Medium-pressure UV lamps that emit strongly below 250 nm take advantage of the fact that the molar absorption coefficient of H_2O_2 is much larger in that wavelength region. However, low-pressure and low-pressure, high-output lamps have a higher overall efficiency and have a distinct advantage if the water strongly absorbs UV in the 200–250 nm region. If the alkalinity is very high (for pH > 7), neither process may be effective because bicarbonate, and particularly carbonate, scavenge •OH radicals. However, with suitable design that accounts for bicarbonate scavenging, this impediment can be overcome.

Many water utilities are now considering the UV/H_2O_2 process (as well as other AOTs such as the O_3/H_2O_2 process) to treat low levels of contaminants, such as pesticides, herbicides, endocrine disruption compounds, pharmaceutical products, and many others. The largest AOT facility is the Orange County Water District's groundwater replenishment system, which is designed to treat NDMA and trace constituents for indirect potable reuse.

The UV/H_2O_2 AOT differs from UV disinfection in that it requires the addition of H_2O_2 and UV doses at least 10–50 times those required for UV disinfection. For example, the UV dose applied in these installations can be >1000 mJ cm^{-2}, which is about 20 times the UV dose required for inactivation of virtually all pathogens. Thus, any drinking water treatment plant that installs a UV/H_2O_2 AOT system will automatically achieve high levels of pathogen inactivation. The UV/H_2O_2 AOT is also used to treat seasonal taste-and-odor problems in drinking water, which may be operated seasonally reducing the overall operating costs of the AOT system.

REFERENCES

APHA, AWWA, and WEF (American Public Health Association, American Water Works Association, and Water Environment Federation). 1998. *Standard Methods for the Examination of Water and Wastewater*. 20th ed. Washington, D.C.: APHA.

Bolton, J.R. 2015. *Ultraviolet Applications Handbook*. 3rd ed. Edmonton, Alta., Canada: ICC Lifelong Learn Inc.

Bolton, J.R.; I. Mayor-Smith; K.G. Linden. 2015. Rethinking the Concepts of Fluence (UV Dose) and Fluence Rate: The Importance of Photon-based Units—A Systemic Review. *Photochemistry and Photobiology*, 91:1252–1262. DOI: 10.1111/php.12512.

Bolton, J.R. and S.R. Cater. 1994. Homogeneous Photodegradation of Pollutants in Contaminated Water: An Introduction. In *Aquatic and Surface Photochemistry*. Edited by G.R. Helz, R.G. Zepp, and D.G. Crosby. Boca Raton, Fla.: Lewis.

Braslavsky, S.E. 2007. Glossary of Terms Used in Photochemistry, Third Edition. *Pure and Applied Chemistry*, 79(3):293–465. DOI:10.1351/pac200779030293.

Collins, J. and J.R. Bolton. 2016. *Advanced Oxidation Handbook*. American Water Works Association, Denver, CO. ISBN 978-1-58321-971-3.

De Latt, J. and M.I. Stefan. 2018. UV Chlorine Process. Chapter 9 in Stefan (2018).

Guerrero, M.A. and R.D. Jones. 1996. Photoinhibition of Marine Nitrifying Bacteria. I. Wavelength-dependent Response. *Marine Ecology Progress Series*, 141:183–192.

IUVA (International Ultraviolet Association). 2021. Far UV-C Radiation: Current State-of Knowledge, IUVA White Paper, IIUVA. https://iuva.org/resources/covid-19/Far%20UV-C%20Radiation-%20Current%20State-of%20Knowledge.pdf.

Legrini, O.; E. Oliveros; A.M. Braun. 1993. Photochemical Processes for Water Treatment. *Chemical Reviews*, 93:671–698.

Olson, R.J. 1981. Differential Photoinhibition of Marine Nitrifying Bacteria: A Possible Mechanism for the Primary Nitrite Maximum. *Journal of Marine Research*, 39(2):227–238.

Oppenländer, T. 2003. *Photochemical Purification of Water and Air: Advanced Oxidation Processes (AOPs)—Principles, Reaction Mechanisms, Reactor Concepts.* New York: Wiley.

Parsons, S., ed. 2004. *Advanced Oxidation Processes for Water and Wastewater Treatment*, London: IWA Publishing.

Ryer, A. 1997. *The Light Measurement Handbook* [Online]. Available: International Light. http://www.intl-lighttech.com/services/light-measurement-handbook.

Stefan, M.I., Ed. 2018. *Advanced Oxidation Processes for Water Treatment: Fundamentals and Applications.* IWA Publishing, London, UK. ISBN 9781780407180.

USEPA. 2006. *Ultraviolet Disinfection Guidance Manual for the Final Long Term 2 Enhanced Surface Water Treatment Rule* [Online]. Available: https://nepis.epa.gov/Exe/ZyPDF.cgi?Dockey=600006T3.txt.

Wayne, R.P. 1988. *Principles and Applications of Photochemistry.* Oxford, UK: Oxford University Press.

White, B. and M. Adams. 2010. Battling Nitrification with Blacklights, *American Water Works Association Opflow*, 36(2):10–13.

Wieringa, F.P. 2006. Five Frequently Asked Questions About UV Safety. *IUVA News,* 8(2):29–32.

Wilczak, A. and H.H. Lai. 2006. Preliminary Bench and Pilot Evaluation of UV-Irradiation for Nitrification Control. In Proc. of the AWWA Annual Conference and Exhibition. Denver, Colo.: AWWA.

CHAPTER 3
Mechanism of UV Disinfection

This chapter explores the process of how absorbed UV photons inactivate microorganisms so effectively. It is important to use the term *inactivate*, rather than *kill* because microorganisms inactivated by UV radiation are typically still metabolically active but cannot replicate.

The process starts with the absorption of UV photons in a given microorganism (First Law of Photochemistry, Chapter 2). Absorption of UV radiation by proteins in the outer cell membranes with high UV doses (fluences) ultimately leads to the disruption of these membranes and hence, the death of the cell arising from leakage of the protoplasm. However, at much lower UV dose (fluence) levels, absorption of UV photons by DNA (or RNA in some viruses) can disrupt the ability of the microorganism to replicate. *Cells that cannot replicate cannot cause infection.*

Most government regulations require that UV equipment inactivate microorganisms to a specific level. For example, Austria and Germany require four orders of magnitude, that is, 4 log (or 99.99 percent) inactivation for all microorganisms. This means that only one microorganism in 10,000 will remain viable. Chapter 4 provides an overview of UV disinfection regulations.

INACTIVATION MECHANISMS

Absorption of UV Radiation by Cell Components

As UV radiation (200–300 nm) passes through a microorganism, it is absorbed by various components in the cell, but only proteins and the nucleotides comprising DNA and RNA absorb a significant amount of UV radiation in this wavelength range. The relative absorbance of proteins versus that of nucleotides in a microorganism is shown in Figure 3-1. It is clear that below 230 nm, proteins absorb most of the UV radiation; however, above 230 nm, absorption is dominated by the

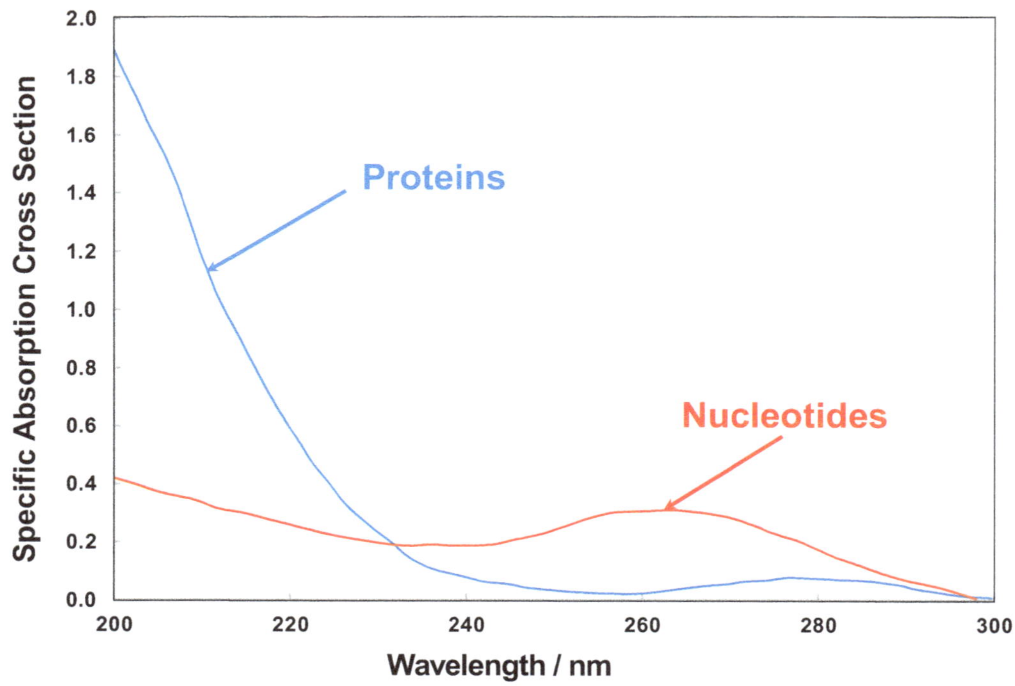

Figure 3-1 Specific absorption cross section versus wavelength for proteins and nucleotides within the cell of a microorganism. Here the dry weight percentages are protein (70%), nucleotides (20%), and other components (10%) (e.g., lipids).*

nucleotides. It should also be noted: the only amino acids that absorb significantly above 210 nm are the aromatic amino acids: tryptophan, phenylalanine, and tyrosine. These three amino acids have a small absorption peak at around 280 nm, with tryptophan being the strongest absorber at 280 nm. For UV disinfection of drinking water, the water itself strongly absorbs radiation below 230 nm, which can make these wavelengths largely unavailable to inactivate the microorganisms, except in the case of some viruses, which have a high germicidal efficiency at wavelengths below 230 nm.*

Another factor to consider is that the overall absorbance of a microorganism is rarely more than 0.02 and usually much less because the optical path length in a cell is so small (< 1 μm). Thus, less than 2 percent of the UV incident on a microorganism is absorbed. Under these conditions, all absorbers behave independently. That is, it is not possible for the protein in the outer membrane of a cell to act as a filter to shield the UV radiation from the nucleus in the center of the cell.

Although UV radiation can be absorbed by proteins (mostly in the outer cell membranes), very high UV doses (fluences) are required, compared to those necessary to disrupt the DNA and RNA. Thus, the important fact is that almost all UV inactivation arises from UV absorption by nucleotides. Therefore, to

* This figure has been constructed by the authors using average spectra for proteins and nucleotides along with the respective mass ratios (see Harm 1980).

understand the mechanism of UV inactivation of microorganisms, the structure and photochemistry of DNA and RNA need to be examined.

Structure and Photochemistry of DNA and RNA

DNA is the fundamental template for life. It is a polymer of nucleotides in a double-stranded helix linked by a sequence of four constituent bases (adenine, cytosine, guanine, and thymine), which constitute the genetic code. These form base pairs (adenine with thymine and cytosine with guanine) held together by hydrogen bonds. This is the glue that binds the two strands of DNA.

RNA has several roles in a cell. Messenger RNA (mRNA) is transcribed from DNA and carries the genetic information from DNA to the ribosomes where proteins are synthesized. Transfer RNA (tRNA) binds with specific amino acids and carries them to the ribosomes for protein synthesis. RNA has four types of nucleotides: adenine, cytosine, and guanine, but uracil replaces thymine.

Most bacteria and protozoa contain DNA in the cell nucleus. Viruses are subcellular entities that can only reproduce by infecting specific host cells. They consist of a DNA or RNA nucleotide helix with a protein outer shell. The DNA or RNA in viruses can be single stranded or double stranded.

All of the nucleotides in DNA or RNA absorb UV radiation between 200–300 nm (see Figure 3-2). Of the four essential nucleotides in DNA, thymine (and in general pyrimidine bases) undergoes a unique photochemical reaction

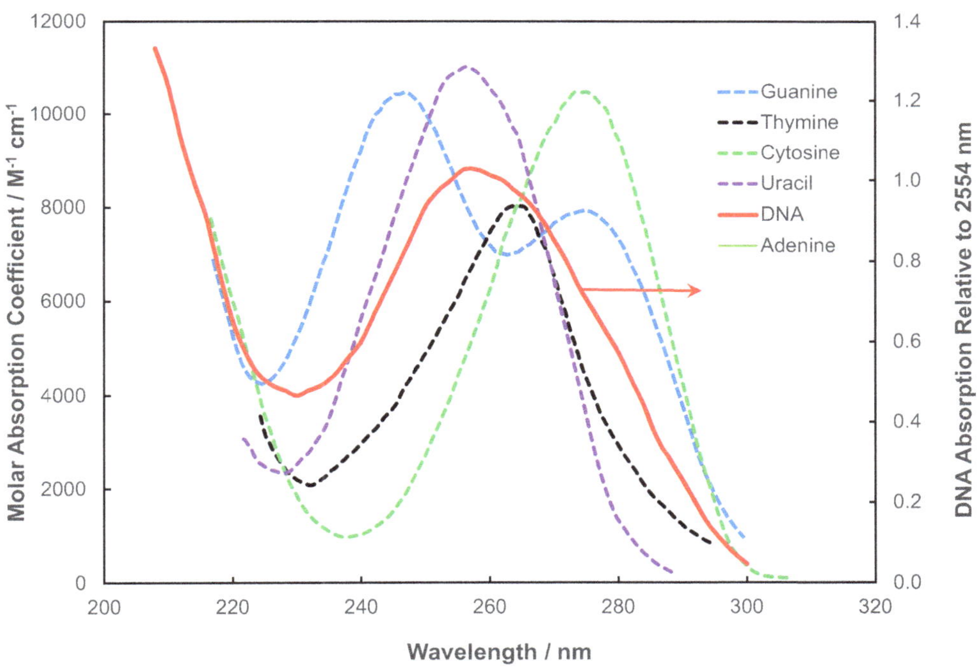

Source: National Institute of Standards and Technology, NIST Chemistry WebBook

Figure 3-2 Absorption spectra of nucleotides and of DNA

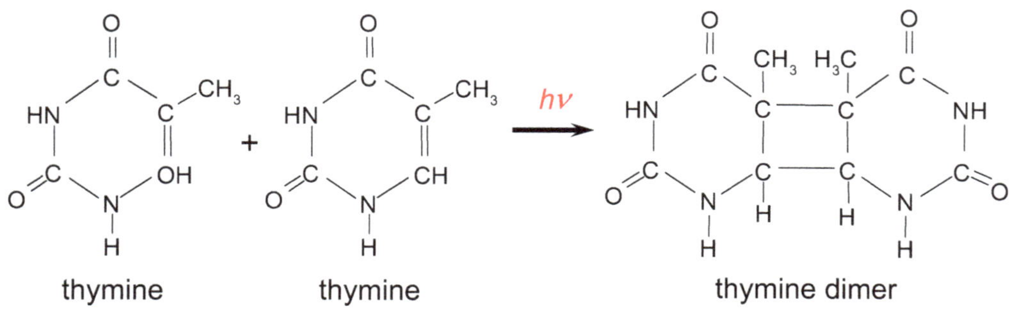

Figure 3-3 Photochemical dimerization of two thymine bases

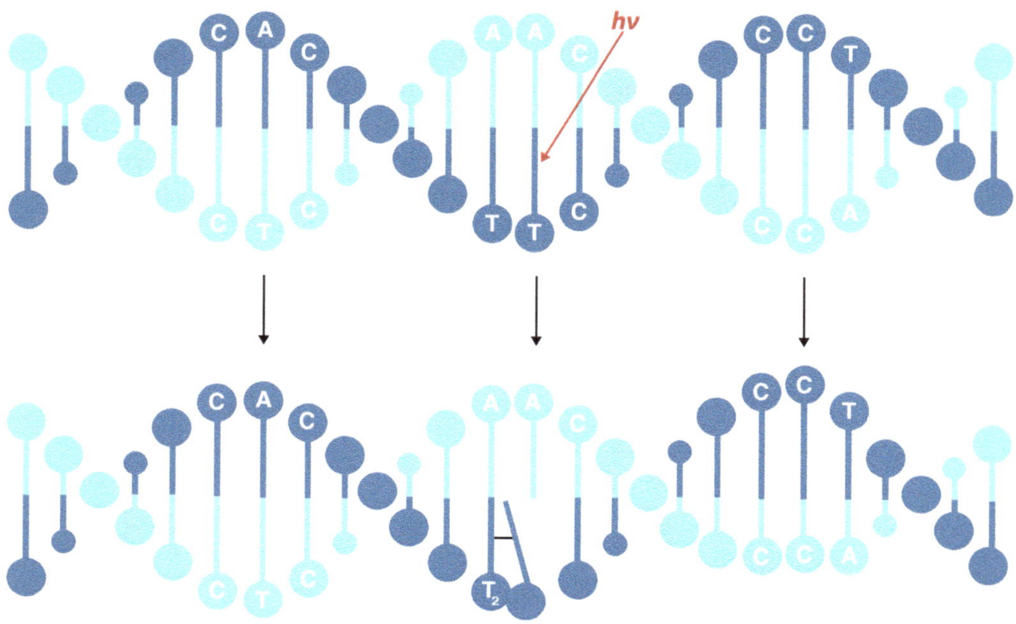

Figure 3-4 Double-stranded DNA chain showing how the formation of thymine dimers disrupts the structure of the chain (A—adenine, G—guanine, T—thymine, C—cytosine)

(see Figure 3-3).† If two thymine bases are adjacent to each other, absorption of a UV photon by one of the thymines leads to a chemical bond between the two thymine molecules (called a thymine dimer).

The formation of thymine dimers (and other similar dimers) disrupts the structure of the DNA so that if a critical number of dimers is formed,‡ the DNA cannot replicate. This prevents infection and is the fundamental mechanism of UV disinfection. Figure 3-4 illustrates the disruption of a DNA chain. For viruses that contain only RNA, a similar photochemical dimerization reaction takes place between two uracil bases.

† Cytosine also forms dimers; however, the quantum yield is much lower than that of thymine.

‡ Oguma et al. (2002) have shown that the generation of approximately 100 thymine dimers is sufficient to inactivate the entire DNA chain.

Viruses can be single- or double-stranded DNA or RNA. There does not appear to be any relation between single and double strands or the presence of DNA or RNA with the UV sensitivity of the virus.

Action Spectra and the Germicidal Factor

The relative inactivation response of a microorganism to UV radiation is determined by the ability of the UV radiation to be absorbed, causing inactivation of the organism through photochemical reactions, such as the dimerization of adjacent thymine pairs in the DNA chains. This relative response is called the *inactivation action spectrum*. Figure 3-5 shows the action spectra of some microorganisms, along with the relative spectrum of DNA. Bolton (2017) has written a review on action spectra. It should be noted that the maximum germicidal effectiveness occurs around 260 nm. The differences in the action spectra probably arise from different compositions of nucleotides in their DNA and RNA.

In many models of the influence of UV radiation on microorganisms, it is necessary to include the relative germicidal effectiveness of the UV radiation at various wavelengths. This is particularly true for polychromatic radiation sources, such as medium-pressure UV lamps. For this purpose, the germicidal factor (GF) has been introduced, such that the GF = 1.00 at 254 nm, the principal wavelength emitted by a low-pressure UV lamp. If the action spectrum for a microorganism is available, it can be normalized to 1.000 at 254 nm. However, in most cases, the action spectra are not known; then the relative absorption spectrum of DNA is often used as the GF. The use of the GF is particularly important for UV

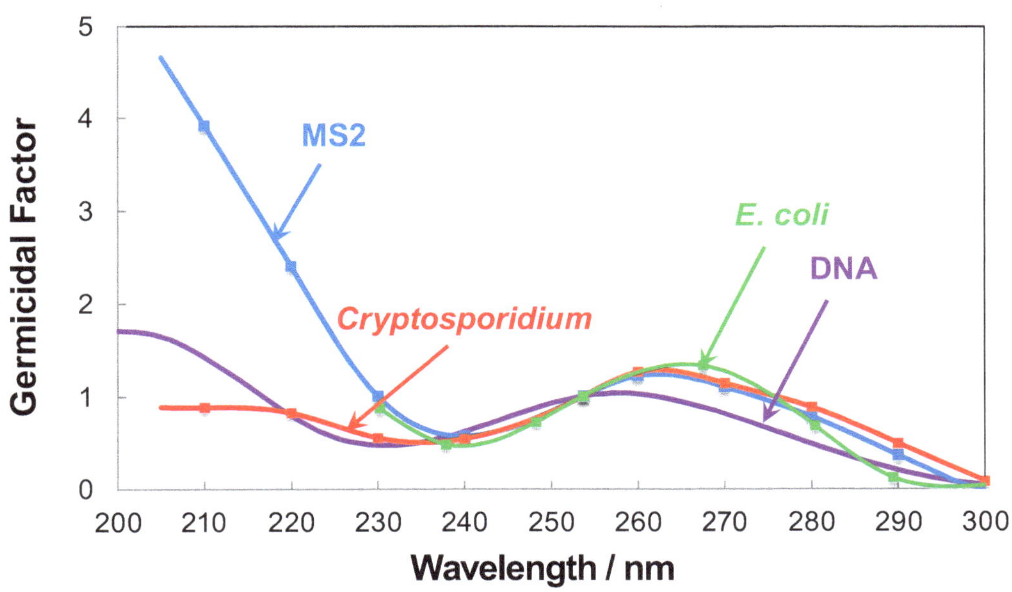

Figure 3-5 Action spectra (relative response versus wavelength) for DNA (——) (von Sonntag and Schuchmann 1992); MS2 colliphage (—■—■—) (Beck et al. 2015); *Cryptosporidium* (—■—■—); (Beck et al. 2015); and *E. coli* (—■—■—) (Gates 1930).

equipment using UV lamps emitting polychromatic UV radiation (e.g., medium-pressure lamps).

Many viruses (e.g., MS2; see Figure 3-5) exhibit a sharply rising GF as the wavelength decreases from 230 to 200 nm. This may arise from excitation energy transfer from proteins in the viral shell to DNA or RNA in the viral core (Chen et al. 2009). Another consideration is that the target pathogen and the validation surrogate have different action spectrum, which could bias the validation results if polychromatic lamps are utilized and is discussed in Chapter 6.

REACTIVATION MECHANISMS

Some microorganisms (particularly bacteria) have mechanisms that can repair or bypass thymine dimer lesions in the DNA chain during the replication process. Some viruses are able to reactivate by using the reactivation enzymes of their host. *Giardia* and *Cryptosporidium* have not yet been shown to reactivate such that they regain infectivity (Linden et al. 2002, Shin et al. 2001, and Oguma et al. 2001). These mechanisms are divided into dark and light mechanisms. The latter are called photoreactivation and are activated by near UV and short-wavelength visible radiation. Harm (1980) has a good discussion of reactivation mechanisms, which is summarized in the following section.

Dark Reactivation Mechanisms

Most bacteria have some reactivation potential in the dark. Various mechanisms have been proposed by which a microorganism can repair DNA damage in the dark:[§]

- *Replacement of UV-damaged nucleotides.* In this case, the UV photoproduct (e.g., thymine dimer) and a sequence of adjacent nucleotides are removed from the DNA and a corresponding sequence is resynthesized.
- *Combination of undamaged regions in replicating DNA molecules.* Undamaged regions of DNA are replicated and reconstituted, so that an intact, double-stranded DNA, identical with the original undamaged DNA, is formed. This is possible because the complete information for a DNA molecule exists totally in each single strand. Thus, one strand can function even if the other strand is damaged.

Photoreactivation Mechanisms

It is possible to have *light-induced reversal of the UV damage*. This is also called *photoenzymatic repair*. This mechanism is the most prevalent reactivation mechanism and was discovered independently in bacteria by Kelner (1949a,1949b) and in bacteriophages by Dulbecco (1949, 1950). Here, a special light-activated

[§] The meaning of the term *dark* in this context means mechanisms that do not require light activation.

enzyme, *photolyase*, splits the thymine dimers, restoring the original structure on the DNA.

If a residual disinfectant (e.g., chlorine or chloramines) is maintained in the distribution system, photoreactivation is not a concern because reactivated microorganisms will be inactivated by the residual disinfectant (with the exception of *Cryptosporidium*).

This process is triggered by the absorption of UVA radiation and is thus called *photoreactivation*. This repair mechanism can be overcome, but this requires a higher UV dose (fluence). This is typically not of concern in water treatment systems that are totally covered; however, if water is stored in open ponds or reservoirs exposed to sunlight, photoreactivation must be considered.

Photoreactivation is not important for treated drinking water, which is rarely exposed to sunlight. But photoreactivation may be particularly important for the UV treatment of wastewater, where the effluent is discharged into a river or lake. Nevertheless, Shafaei et al. (2016) showed that a UV dose of >20 mJ cm^{-2} applied to wastewater totally shuts down photoreactivation. Since most wastewater treatment plants operate at UV doses much higher than 20 mJ cm^{-2}, photoreactivation is not of significant concern.

UV SENSITIVITY OF MICROORGANISMS

Microorganisms exhibit a wide variation in their sensitivity to UV radiation. The measurement of the UV sensitivity of microorganisms at a specific wavelength, or narrow band of wavelengths, is performed through the determination of a UV dose (fluence)–response curve.

Collimated Beam Tests

The UV sensitivity of microorganisms is determined using a *collimated beam test*, where a sample suspension of the microorganisms[¶] is exposed to UV radiation (usually in an open Petri dish) in a collimated beam apparatus where the sample is exposed to beams of UV radiation that are almost parallel.[**] Figure 3-6 illustrates two types of collimated beam apparatus. Such an apparatus can be constructed from commonly available materials or purchased from a vendor.

Bolton and Linden (2003) initially established a standard protocol for these measurements. This was later expanded into a full guidance Protocol by the International Ultraviolet Association (IUVA 2015). The Protocol specifies that the average irradiance E(avg) in the Petri dish is given by

$$E(\text{avg}) = E^0 \times \text{PF} \times \text{DF} \times \text{RF} \times \text{WF} \qquad \text{(Eq 3-1)}$$

¶ Hereinafter, this will be referred to as the *suspension*.

** The term *collimated beam apparatus* is widely used, but it is important to realize that the beam is not truly collimated, in the sense that all rays are parallel. The beam does diverge, and thus, there is a need for a divergence factor. Often, the term *quasi* collimated beam is used to reflect this reality.

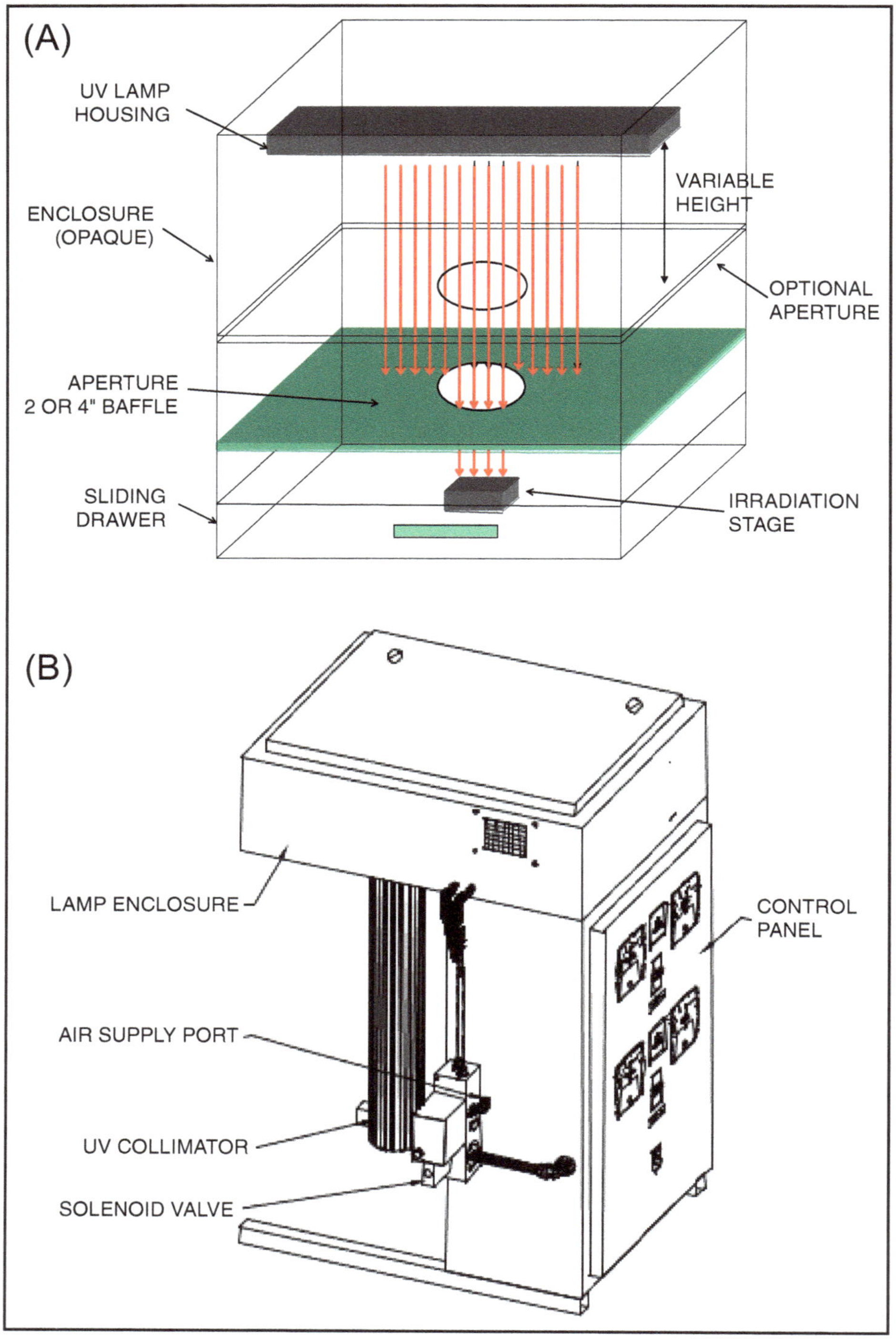

Figure 3-6 Home-made collimated beam apparatus (A: courtesy of Prof. Linden, personal communication); Commercial collimated beam apparatus (B: courtesy of Calgon Carbon Corporation, Pittsburgh, Pa.)

where E^o (usually measured with a calibrated radiometer) is the incident irradiance (mW cm^{-2}) at the center of the Petri dish at the level of the top surface of the water in the dish; PF is the Petri factor, which is the ratio of the average incident irradiance across the Petri dish divided by E^o; DF is the Divergence factor, defined as $D/(D + l)$, where D is the distance from the top surface of the water in the dish to the center of the UV lamp and l (cm) is the suspension depth; RF is the Reflection factor (0.975 at 254 nm); and WF is the water factor, defined as $(1 - al)/[al\ln(10)]$, where a is the absorption coefficient (cm^{-1}) of the suspension.

In a properly conducted determination of the UV dose (fluence)–response curve for a microorganism,[††] where the sample is well stirred and the above corrections are made, an accurate determination of the average irradiance E(avg) in the water suspension can be made from the incident irradiance. The product of E(avg) and the exposure time in seconds gives the delivered UV dose (fluence) F. For each exposure time, the sample suspension is sent to a microbiological laboratory where the concentration of viable organisms is determined.

UV Dose (Fluence)–Response Curves

Figure 3-7 shows an example of a UV dose (fluence)–response curve. The UV dose (fluence)–response curve can exhibit up to three phases. For many bacteria, there exists a *shoulder* phase, such as that seen for *B. subtilis* (see Figure 3-7), where $\log(N/N_0)$ changes only slowly with the UV dose (fluence) at the beginning. This is followed by a *semilog* phase, where $\log(N/N_0)$ decreases linearly with UV dose (fluence). Finally, there may exist a *saturation* phase, in which $\log(N/N_0)$ changes only slowly with UV dose (fluence).

The *shoulder* phase is thought to be associated with DNA repair mechanisms, or it may arise from the requirement of several *hits* (DNA damage sites) before a microorganism becomes inactivated (Roessler and Severin 1996). The saturation phase may arise from several causes, such as microorganisms clumping or associating with particles (this is particularly prevalent in wastewater systems).

The *semilog* phase is characterized by the rate equation

$$\log\left(\frac{N}{N_0}\right) = -kF \qquad \text{(Eq 3-2)}$$

where k is a standard fluence-based rate constant [$cm^2\ mJ^{-1}$ or m^2J^{-1}] that is characteristic of the microorganism, and F is the UV dose (fluence). Other more sophisticated equations have been developed to model the other two phases (Roessler and Severin 1996).

[††] Unfortunately, until recently most response curves were based on the incident irradiance, rather than the average irradiance in the microorganism suspension. Thus, a significant part of the published literature should be considered as upper limits because the irradiance has been overestimated.

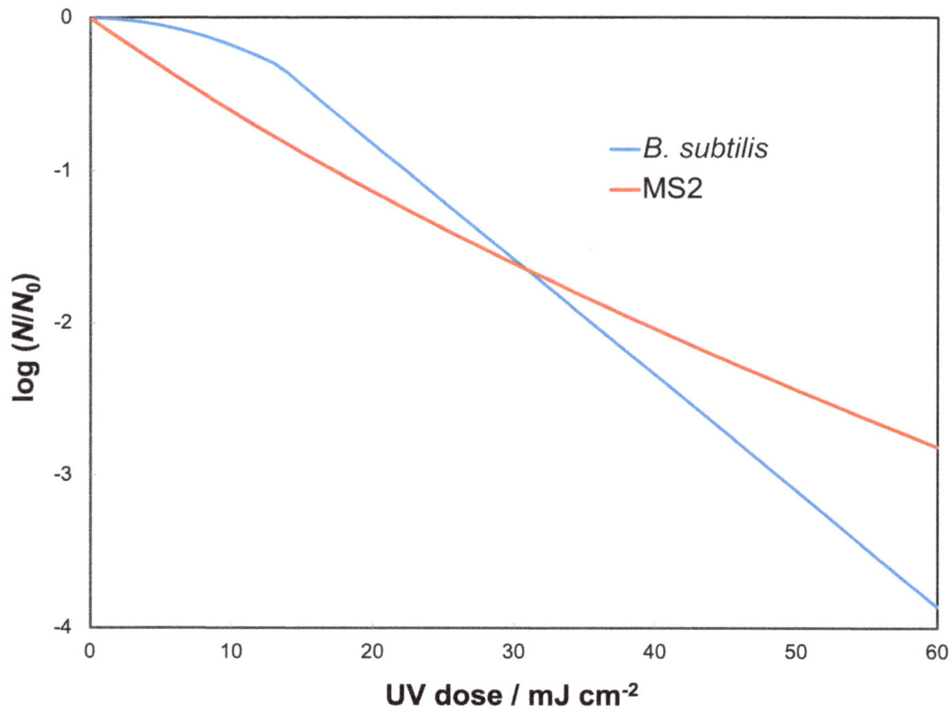

Figure 3-7 UV dose–response curves for *B. subtilis* spores (ÖNORM 2020) and the virus MS2 coliphage (USEPA 2020)

Relative UV Sensitivity of Pathogenic Microorganisms

The UV sensitivity of pathogenic microorganisms varies considerably. The currently accepted order is

bacteria ≈ protozoa > most viruses > bacterial spores > adenovirus > algae
(most sensitive)——————————————(least sensitive) (Eq 3-3)

Adenovirus is listed separately because it is unique among waterborne viruses because it is the least sensitive to LP UV radiation of any virus.‡‡ The reason for this insensitivity may arise from the fact that adenovirus can repair UV damage to its double-stranded DNA in the host cells (Beck et al. 2014, 2018).

Tables 3-1a, 3-1b, and 3-1c present a compilation of data on the UV sensitivity of numerous pathogenic microorganisms, as determined in a bench-scale collimated beam apparatus. The UV doses (fluences) listed are for 4-log (99.99%) inactivation. Note that most bacteria are capable of photoreactivation and thus require a higher UV dose (fluence) for inactivation when near-UV and visible light are present.

‡‡ This apparently is only true for LP UV radiation (254 nm). Linden et al. (2007) have presented evidence that when an MP UV lamp is used, the UV dose (fluence) required for 4-log inactivation of adenovirus can be as low as 60 mJ cm^{-2}.

Table 3-1a Bench-scale UV dose (fluence) at 254 nm required for 4-log (99.99%) inactivation of various bacteria*,†

Bacterium	UV Dose (Fluence) (mJ cm^{-2}) for 4-Log (99.99%) Inactivation	
	Without Photoreactivation	With Photoreactivation‡
Aeromonas hydrophila	5.5	NA
Ammonia oxidizing bacteria	15-20 (3.5 log inactivation)	NA
Bacillus anthracis spores	>60	—
Bacillus subtilis spores	62	—
Campylobacter jejuni	4.6	NA
Citrobacter diversus	11	NA
Clostridium perfringens	23.5	NA
Corynebacterium diphtheria	6	NA
Coxiella bumetti	3	NA
Enterobacter cloacae	10	33
Enterocolitica faecium	17	20
Escherichia coli ATCC 11229	7.6	28
Escherichia coli O157:H7	7	25
Escherichia coli wild type	8	NA
Klebsiella pneumoniae	12	31
Klebsiella terrigena	12	NA
Legionella pneumophila	6	NA
Mycobacterium smegmatis	20	27
Mycobacterium tuberculosis	20	NA
Pseuodomonas aeruginosa	9	19
Pseudomonas mallei	15	NA
Salmonella anatum (from human feces)	7	NA
Salmonella enteritidis (from human feces)	10	NA
Salmonella typhi ATCC 19430	8.2	NA
Salmonella typhimurium (from human feces)	7	NA
Salmonella marcescens	13	30
Shigella dysenteriae ATTC 29027	3	NA
Shigella sonnei ATTC 9290	8	NA
Staphylococcus aureus	10	NA

Table 3-1a Bench-scale UV dose (fluence) at 254 nm required for 4-log (99.99%) inactivation of various bacteria*,† (*continued*)

Bacterium	UV Dose (Fluence) (mJ cm^{-2}) for 4-Log (99.99%) Inactivation	
	Without Photoreactivation	With Photoreactivation‡
Streptococcus faecalis ATTC 29212	11	NA
Vibrio cholerae ATTC 25872	2.8	21
Yersinia enterocolitica ATTC 27729	5	NA

* Data extracted from Hoyer (1998) and Masjoudi et al. (2021).

† The UV doses (fluences) listed in this table were obtained at the bench scale with UV exposure at 254 nm, and are approximate for the bacterium listed. The UV doses (fluences) do vary somewhat among various strains of a given bacterium.

‡ "With photoreactivation" means that the cell suspension has been exposed to visible and/or near UV light after the UV exposure.

Table 3-1b Bench-scale UV dose (fluence) at 254 nm required for 4-log (99.99%) inactivation of various viruses*,†

Virus	Classification	UV Dose (Fluence) (mJ cm^{-2}) for 4-Log (99.99%) Inactivation	
		Without Photoreactivation	With Photoreactivation‡
Adenovirus Type 40	dsDNA	130§	—
Adenovirus Type 41	dsDNA	222§	—
B40-8 phage		29	—
Coxsackievirus B5	ssRNA	36	—
Hepatitis A HM175	ssRNA	16	—
MS2 bacteriophage	ssRNA	96	—
Poliovirus Type 1	ssRNA	31	—
PRD-1 phage		33	—
FX174 phage	ssDNA	9	
Rotavirus SA11	dsRNA	38	—
SARS-CoV2	ssRNA	4	—
S. aureus phage A994		38	—

* Data extracted from Hoyer (1998) and Masjoudi et al. (2021).

† The UV doses (fluences) listed in this table were obtained at the bench scale and are approximate for the virus listed. The UV doses (fluences) do vary somewhat among various strains of a given virus.

‡ It has been found that generally viruses do not undergo photoreactivation

§ The USEPA has set a 4-log inactivation UV dose of 186 mJ cm^{-2} as the virus limit, based on the inactivation of adenovirus but taking into account an 80 percent credible interval.

Table 3-1c Bench-scale UV dose (fluence) at 254 nm required for 4-log (99.99%) inactivation of various protozoa*,†

Protozoa	UV Dose (Fluence) ($mJ\ cm^{-2}$) for 4-Log (99.99%) Inactivation	
	Without Photoreactivation	With Photoreactivation‡
Giardia lamblia	<10	—
Giardia muris§	<10	—
Cryptosporidium parvum	<10	—

* Data extracted from Clancy et al. (2000), Craik et al. (2000, 2001), and Masjoudi et al. (2021).
† "With photoreactivation" means that the cell suspension as has been exposed to visible and/or near UV light after the UV exposure.
‡ It is generally not known whether or not these specific protozoa undergo photoreactivation.
§ Not considered to be a human pathogen.

Protozoa represent a class of pathogens (much larger in size than bacteria) that have become a threat to distributed water systems. The two most important pathogens are *Giardia lamblia* (causing giardiasis) and *Cryptosporidium parvum* (causing cryptosporidiosis). These pathogens are regulated by USEPA (as described in Chapter 4).

Giardia lamblia

Giardia lamblia can infect the intestinal tract of most fish, reptiles, birds, and mammals. *Giardia* cysts are quite large (about 10–15-μm-long and oval in shape) compared to bacteria and are often found in rivers and streams exposed to wildlife; for this reason, the gastrointestinal disease giardiasis is commonly referred to as *beaver fever*.

Cryptosporidium parvum

Cryptosporidium parvum can infect the intestinal tract of most fish, reptiles, birds, and mammals. When ingested, *Cryptosporidium* oocysts (about 3–5 μm in diameter) cause mild to severe diarrhea but can be fatal to immunocompromised individuals, such as those with HIV/AIDS. Large-scale outbreaks have been traced to water contaminated by agricultural runoff, particularly from large-scale livestock operations.

Since the major outbreak of cryptosporidiosis in Milwaukee, Wis. in 1993, where more than 400,000 persons were infected, *Cryptosporidium* has become the focal point of the USEPA surface water regulations (LT2ESWTR) (USEPA 2006).

UV Sensitivity of Protozoa

Before 1998, protozoa would have been placed well to the right in Eq 3-2 because it was widely perceived that protozoa (e.g., *Cryptosporidium* or *Giardia*) were very insensitive to UV radiation. This perception arose because the assays used to assess the viability of protozoa were based on UV damage to the cell membranes rather than on an infectivity assay. The following from the USEPA illustrates this widely held perception (USEPA 1996):

> *Ultraviolet light is an effective disinfection agent for drinking water at sufficient intensity and appropriate wavelength and exposure time. It is suitable for use on clean water sources more than it is for suspended matter–containing water. The order of disinfection resistance with few exceptions is bacteria (least resistant) < viruses < bacterial spores < protozoan oocysts (most resistant).*

In 1997 and 1998, a unique team of photochemists, photobiologists, and parasitologists worked together to conduct a series of experiments on *Cryptosporidium*, involving an in vivo assay, in which oocysts exposed to UV radiation were fed to neonatal mice. The advantage of this assay was that it provided a test that was focused on the ability of UV-treated oocysts to cause an infection in the mice. The results were remarkable in that almost no mice were infected by UV-treated oocysts even at very low UV doses (fluences). A paper describing this work was presented in June, 1998 (Bolton et al. 1998) announcing the new results (see also Bukhari et al. 1999). Research from many laboratories (Clancy et al. 2000; Craik et al. 2000, 2001; Linden et al. 2001; Mofidi et al. 2001; Shin et al. 2001; Rochelle et al. 2002; Morita et al. 2002) has confirmed that *Cryptosporidium* and *Giardia* can be inactivated by more than 4 logs at UV doses (fluences) as low as 10 mJ cm^{-2} or less.

As a result of this research, the attitude of the USEPA changed markedly such that in the new LT2ESWTR regulations (USEPA 2006), UV disinfection is treated as the best available technology.

UV DOSE (FLUENCE) IN A UV REACTOR

It is important to realize that at the bench scale in a collimated beam apparatus, all the microorganisms receive essentially the same UV dose (fluence) because the suspension is well stirred. This is not the case in a UV reactor where microorganisms can traverse a variety of paths through the reactor. The total UV dose (fluence) experienced by each individual microorganism is the product of the irradiance (fluence rate) in each volume element traversed times the transit time over that volume element in seconds integrated over the actual path taken by the microorganism. Because each path is different, the UV doses (fluences) received by the various microorganisms are different. This results in a UV dose (fluence) distribution (see Figure 3-8). The dose distribution concept is further illustrated in Chapter 6 (Validation of UV Reactor Performance).

Using a mathematical model for the irradiance (fluence rate) distribution in the UV reactor, one can calculate the volume averaged irradiance (fluence rate). Multiplication of this value by the hydraulic residence time in seconds yields the *maximum theoretical UV dose (fluence)* $F_{ideal}(D_{ideal})$ for that reactor under specified conditions of flow rate, UVT, and lamp power. An *ideal* UV reactor,

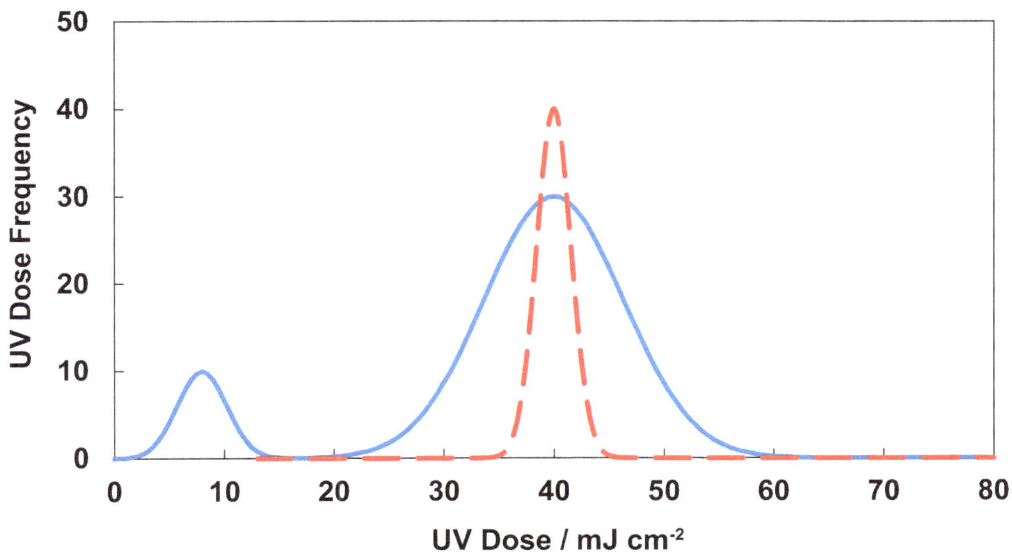

Figure 3-8 Two possible UV dose distributions. The red curve indicates a relatively uniform distribution, whereas the blue curve indicates regions of very low UV dose.

with perfect radial mixing, would have a UV dose (fluence) distribution with an extremely narrow UV dose (fluence) distribution, approximately the value F_{ideal} (D_{ideal}). However, real UV reactors have UV dose (fluence) distributions that are broader (the breadth is a qualitative measure of the mixing efficiency). Figure 3-8 illustrates two distributions, one of which is much broader than the other. Even though each UV reactor may have the same value of F_{ideal}, the real delivered UV dose (fluence) will be lower than $F_{\text{ideal}}(D_{\text{ideal}})$ for the UV reactor with the narrower UV dose (fluence) distribution (dashed curve in Figure 3-8) and much lower than $F_{\text{ideal}}(D_{\text{ideal}})$ for the UV reactor with the broader and more complex UV dose (fluence) distribution (solid curve in Figure 3-8). In the case of the latter distribution, the situation is even worse because of the separate distribution at low UV doses. This will severely limit the ability of this UV reactor to achieve high UV doses and hence high inactivation levels.

It is important to understand that in any real UV reactor, microorganisms flowing through the reactor will receive a variety of UV doses (fluences), some higher and some lower than the mean UV dose (fluence). The microorganisms that receive a low UV dose (fluence) are of particular concern (e.g., the lower peak in Figure 3-8) and can limit the ability of the UV reactor to deliver the required log inactivation.

The method currently used to assess the average UV dose (fluence) is to carry out a biodosimetry test (see Chapter 6), which yields the reduction equivalent dose (RED). If the UV dose (fluence) distribution can be obtained (e.g., from computational fluid dynamics models, see Chapter 6), the RED can be estimated by convoluting the UV dose (fluence) distribution with the UV sensitivity kinetics of the test microorganism. Chapter 6 provides a more detailed description of the RED concept and calculations.

FAR-UV DISINFECTION

There is now considerable interest in UV disinfection in the *Far-UV* region (200–225 nm). The concept is that many pathogens have an enhanced UV sensitivity in this region, and furthermore UV in this region is absorbed very strongly in the upper layer of the skin, which sloughs off every day. Some manufacturers are marketing UV disinfection systems that they claim are safe for human exposure. The UV radiation source is usually a Kr-Cl excilamp at 222 nm, filtered to remove longer wavelengths in the emission. Almost all systems developed are intended for open air UV disinfection. Far-UV applications in water would be limited because of high absorption in the far-UV region. The International Ultraviolet Association has written a white paper on Far-UV (IUVA 2021).

REFERENCES

Beck, S.E.; R.A. Rodriguez; K.G. Linden; T.M. Hargy; T.C. Larason; H. B. Wright. 2014. Wavelength Dependent UV Inactivation and DNA Damage of Adenovirus as Measured by Cell Culture Infectivity and Long Range Quantitative PCR, *Environmental Science and Technology*, 48(1):591-598. DOI.org/10.1021/es403850b.

Beck, S.E.; H.B. Wright; T.M. Hargy; T.C. Larason; K.G. Linden. 2015. Action Spectra for Validation of Pathogen Disinfection in Medium-pressure Ultraviolet (UV) systems, *Water Research*, 70:27–37. DOI.org/10.1016/j.watres.2014.11.028.

Beck, S.E.; N.M. Hull; C. Poepping; K.G. Linden. 2018. Wavelength-Dependent Damage to Adenoviral Proteins Across the Germicidal UV Spectrum, *Environmental Science and Technology*, 52:223–229. DOI: 10.1021/acs.est.7b04602.

Bolton, J.R. 2017. Action Spectra: A Review. *IUVA News*, 19(2):10–12.

Bolton, J.R.; B. Dussert; Z. Bukhari; T.M. Hargy; J.L. Clancy. 1998. Inactivation of *Cryptosporidium parvum* by Medium-Pressure Ultraviolet Light in Finished Drinking Water. In Proc. of the AWWA Annual Conference. Denver, Colo.: AWWA.

Bolton, J.R. and K.G. Linden. 2003. Standardization of Methods for Fluence (UV Dose) Determination in Bench-Scale UV Experiments. *Journal of Environmental Engineering*, 129(3):209–215.

Bukhari, Z.; T.M. Hargy; J.R. Bolton; B. Dussert; J.L. Clancy. 1999. Medium-Pressure UV Light for Oocyst Inactivation. *Journal AWWA*, 91(3):86–94.

Chen, R.-Z.; S.A. Craik; J.R. Bolton. 2009. Comparison of the Action Spectra of *Bacillus subtilis* Spores and Relative DNA Absorbance Spectra for the Determination of Germicidal Fluence (UV dose) in the Ultraviolet Disinfection of Water. *Water Research*, 43:5087–5096. doi:10.1016/j.watres.2009.08.032.

Clancy, J.L.; Z. Bukhari; T.M. Hargy; J.R. Bolton; B. Dussert; M.M. Marshall. 2000. Using UV to Inactivate *Cryptosporidium*. *Journal AWWA*, 92(9):97–104.

Craik, S.A.; G.R. Finch; J.R. Bolton; M. Belosevic. 2000. Inactivation of *Giardia muris* Cysts Using Medium-Pressure Ultraviolet Irradiation in Filtered Drinking Water. *Water Research*, 34(18):4325–4332.

Craik, S.A.; D. Weldon; G.R. Finch; J.R. Bolton; M. Belosevic. 2001. Inactivation of *Cryptosporidium parvum* Oocysts Using Medium- and Low-Pressure Ultraviolet Radiation. *Water Research*, 35(6):1387–1398.

Dulbecco, R. 1950. Experiments on Photoreactivation of Bacteriophages Inactivated With Ultraviolet Radiation. *Journal of Bacteriology*, 59:329–347.

Dulbecco, R. 1949. Reactivation of Ultraviolet-Inactivated Bacteriophage by Visible Light. *Nature*, 163:949–950.

Gates, F.L. 1930. A Study of the Bactericidal Action of Ultraviolet Light III: The Absorption of Ultraviolet Light by Bacteria. *Journal of General Physiology*, 14:31–42.

Harm, W. 1980. *Biological Effects of Ultraviolet Radiation*. Cambridge, UK: Cambridge University Press.

Hoyer, O. 1998. Testing Performance and Monitoring of UV Systems for Drinking Water Disinfection. *Water Supply*, 16(1/2):419–424.

IUVA. 2021. Far-UV Radiation: Current State-of Knowledge, International Ultraviolet Association (IUVA), Chevy Chase, MD. https://iuva.org/resources/covid-19/Far%20UV-C%20Radiation-%20Current%20State-of%20Knowledge.pdf.

IUVA. 2015. Protocol for the Determination of Fluence (UV Dose) Using A Low-Pressure or Low-Pressure High-Output UV Lamp in Bench-Scale Collimated Beam Ultraviolet Experiments, International Ultraviolet Association. https://iuva.org/Guidance-Documents.

Kelner, A. 1949a. Effect of Visible Light on the Recovery of *Streptomyces griseus condidia* From Ultraviolet Irradiation Injury. Proceedings of the National Academy of the United States of America, 35(2):73–79.

Kelner, A. 1949b. Photoreactivation of Ultraviolet-Irradiated *Escherichia coli*, With Special Reference to the Dose–Reduction Principle and to Ultraviolet-Induced Mutation. *Journal of Bacteriology*, 58:511–522.

Linden, K.G.; G.-A. Shin; G. Faubert; W. Cairns; M.D. Sobsey. 2002. UV Disinfection of *Guaria lamblia* Cysts in Water, *Environmental Science and Technology*, 36:2519–2522.

Linden, K.G.; G.-A. Shin; M.D. Sobsey. 2001. Comparative Effectiveness of UV Wavelengths for the Inactivation of *Cryptosporidium parvum* Oocysts in Water. *Water Science and Technology*, 43(12):171–174.

Linden, K.G.; J. Thurston; R. Schaefer; J.P. Malley, Jr. 2007. Enhanced UV Inactivation of Adenoviruses Under Polychromatic UV Lamps. *Applied and Environmental Microbiology*, 73(23):7571–7574. DOI:10.1128/AEM.01587-07.

Masjoudi, M.; M. Mohseni; J.R. Bolton. 2021. Fluence (UV Dose) Required to Achieve Incremental Log Inactivation of Bacteria, Protozoa, Viruses and Algae. *Journal of Research of the National Institute of Standards and Technology*, 126:126021. DOI.org/10.6028/jres.126.021.

Mofidi, A.A.; H. Baribeau; P.A. Rochelle; R. De Leon; B.M. Coffey; J.F. Green. 2001. Disinfection of *Cryptosporidium parvum* with Polychromatic UV Light. *Journal AWWA*, 93(6):95–109.

Morita, S.; A. Namikoshi; T. Hirata; K. Oguma; H. Katayama; S. Ohgaki; N. Motoyama; M. Fujiwara. 2002. Efficacy of UV Irradiation in Inactivating *Cryptosporidium parvum* Oocysts. *Applied and Environmental Microbiology*, 68(11):5387–5393.

Oguma K.; H. Katayama; H. Mitani; S. Morita; T. Hirata; S. Ohgaki. 2001. Determination of Pyrimidine Dimers in *Escherichia coli* and *Cryptosporidium parvum* During UV Light Inactivation, Photoreactivation, and Dark Repair. *Applied and Environmental Microbiology*, 67:4630–4637. DOI: 10.1128/AEM.67.10.4630–4637.2001.

Oguma, K.; H. Katayama; S. Ohgaki. 2002. Photoreactivation of *Escherichia coli* After Low- or Medium-Pressure UV Disinfection Determined by an Endonuclease Sensitive Site Assay. *Applied and Environmental Microbiology*, 68(12):6029–6035. DOI: 10.1128/AEM.68.12.6029–6035.2002.

ÖNORM M 5873-1. 2020. *Devices for Disinfecting Water using Ultraviolet Radiation. Part 1: Devices with UV low pressure lamps*. Austrian Standards International, Vienna, Austria. (in German) https://shop.austrian-standards.at/action/de/public/details/667583/OENORM_M_5873-1_2020_01_01.

Rochelle, P.A.; M.M. Marshall; J.R. Mead; A.M. Johnson; D.G. Korich; J.S. Rosen; R. De Leon. 2002. Comparison of In Vitro Cell Culture and a Mouse Assay for Measuring Infectivity of *Cryptosporidium parvum. Applied and Environmental Microbiology*, 68:3809–3817. DOI: 10.1128/AEM.68.8.3809–3817.2002.

Roessler, P.F. and B.F. Severin. 1996. Ultraviolet Light Disinfection of Water and Wastewater. In *Modeling Disease Transmission and Its Prevention by Disinfection*. Edited by C.J. Hurst. Cambridge, UK: Cambridge University Press.

Shafaei, S.; J.R. Bolton; M. Gamel El-Din. 2016. Effect of Environmental Factors on Photoreactivation of Microorganisms under Indoor Conditions. *International Journal of Environmental, Chemical, Ecological, Geological and Geophysical Engineering*, 10(5):468–471.

Shin, G.-A.; K.G. Linden; M.J. Arrowood; M.D. Sobsey. 2001. Low-pressure UV Inactivation and DNA Repair Potential of *Cryptosporidium parvum Oocysts, Applied and Environmental Microbiology*. 67:3029–3032. DOI: 10.1128/AEM.67.7.3029–3032.2001.

USEPA. 2020. Wright, H., T. Brooks and M. Heath. 2020. *Innovative Approaches for Validation of Ultraviolet Disinfection Reactors for Drinking Water Systems*. Report EPA/600/R20/094, US Environmental Protection Agency, Washington, DC. https://cfpub.epa.gov/si/si_public_record_report.cfm?Lab=CESER&count=10000&dirEntryId=349759&searchall=&showcriteria=2&simplesearch=0&timstype=.

USEPA. 2006. Long Term 2 Enhanced Surface Water Treatment Rule (LT2) [Online]. Available: https://www.epa.gov/dwreginfo/long-term-2-enhanced-surface-water-treatment-rule-documents.

USEPA. 1996. *Ultraviolet Light Disinfection Technology in Drinking Water Application—An Overview*. Report No. 811-R-96-002. Washington, D.C: USEPA.

von Sonntag, C. and H.-P. Schuchmann. 1992. UV Disinfection of Drinking Water and Byproduct Formation—Some Basic Considerations. *Journal of Water Supply: Research and Technology—AQUA*, 41(2):67–74.

CHAPTER 4
UV Disinfection Regulations

This chapter primarily describes the USEPA regulations that recognize UV disinfection as a treatment technique to meet public health requirements. Currently, three regulations specifically include UV disinfection as a compliance technology: LT2ESWTR (USEPA 2006a), the Stage 2 DBPR (USEPA 2006b), and the GWR (USEPA 2006c). Other applications of UV disinfection, such as reuse applications, do not currently have USEPA requirements or recommendations; therefore, these applications are not described in this chapter.

In addition, there are regulations in other countries that regulate UV disinfection of drinking water. For example, the Deutscher Verein des Gas und Wasserfaches (German Association for Gas and Water) (ÖNORM and DVGW 2020) regulates UV disinfection and validation practices in Germany and Austria. The ÖNORM and DVGW standards are referred to and sometimes used in North America. Austria and Germany have now adopted joint UV standards (DIN 2020). However, these regulations are not discussed in this chapter because USEPA standards will regulate UV disinfection application in the United States. The NSF International has a standard for UV disinfection for small flows (i.e., NSF Standard 55) (NSF 2020); however, these standards may not meet USEPA requirements and are not discussed in this chapter.

The federal government in Canada has issued a set of drinking water guidelines (Health Canada 2020); however, in Canada drinking water regulations are set by each province or territory and are not described in this chapter.

REGULATORY BACKGROUND

The USEPA's SWTR provides minimum disinfection requirements for water utilities using surface water or groundwater under the direct influence (GWUDI) of surface water. The focus of SWTR is on the removal/inactivation of *Giardia* cysts and viruses. The Interim Enhanced Surface Water Treatment Rule (IESWTR) and

Long Term 1 ESWTR (LT1ESWTR) built upon SWTR requirements by establishing baseline treatment requirements for *Cryptosporidium* for filtered systems.

The LT2ESWTR supplements these regulations by establishing additional treatment technique requirements for water utilities with greater vulnerability to *Cryptosporidium*. The LT2ESWTR was implemented simultaneously with the Stage 2 DBPR. Table 4-1 summarizes the requirements for these microbial and disinfection byproduct rules.

1. The requirements for filtered water utilities are in addition to removal achieved by conventional treatment complying with the IESWTR and LT1ESWTR.
2. Specific requirements for each plant depend on source water monitoring results.
3. The Stage 2 DBPR bases compliance on total trihalomethane (TTHM) and five haloacetic acid (HAA5) levels at locations identified in the initial distribution system evaluation. Compliance is based on the locational running annual averages (LRAAs) for those locations that must not exceed the maximum contaminant levels for TTHMs and HAA5s.

Table 4-1 Summary of microbial and disinfection byproduct rules (USEPA 2006a, 2006b, 2006c)

Surface Water Treatment Rules – Minimum Treatment Requirements			
Regulation	***Giardia***	**Virus**	***Cryptosporidium***
SWTR	3 log removal and/or inactivation	4 log removal and/or inactivation	Not addressed
IESWTR and LT1ESWTR	No change from SWTR		2 log removal
LT2ESWTR	No change from SWTR		0–2.5 log additional treatment for filtered water utilities
			2–3 log treatment for unfiltered water utilities

Disinfection Byproduct Rules – Maximum Contaminant Levels (MCLs)					
Regulation	**Compliance**	**Total Trihalomethanes (TTHM) (μg/L)**	**Five Haloacetic Acids (HAA5) (μg/L)**	**Bromate (μg/L)**	**Chlorite (μg/L)**
Stage 1 DBPR	Running Annual Average	80	60	10	1000
Stage 2 DBPR	Locational Running Annual Average	80	60	No change from Stage 1	

UV disinfection could be used as a part of a compliance strategy for the Stage 2 DBPR by replacing the primary disinfectant with UV disinfection to achieve *Giardia* or *Cryptosporidium* inactivation. Virus credits can be achieved with short contact time chlorine treatment, and chloramines would be needed for the secondary disinfectant. The Stage 2 DBPR is not described in detail in this chapter because it defers to the LT2ESWTR requirements when UV disinfection is used.

LONG-TERM 2 ENHANCED SURFACE WATER TREATMENT RULE

The LT2ESWTR was developed through the negotiations of a federal advisory committee, which reached a consensus agreement in principle in September 2000. In essence, the LT2ESWTR requires additional treatment for *Cryptosporidium* by all unfiltered water utilities and by a subset of filtered water utilities with elevated source water *Cryptosporidium* levels, as demonstrated through monitoring. The LT2ESWTR was proposed in August 2003, and the final LT2ESWTR was published in the *Federal Register* on January 5, 2006 (USEPA 2006). UV disinfection is one of the technologies recognized for inactivation of *Cryptosporidium* and is one of the few technologies accepted for high levels (2 to 3 log) of *Cryptosporidium* inactivation. Accordingly, there are specific requirements related to UV disinfection in the LT2ESWTR.

Along with requiring the addition of *Cryptosporidium* inactivation and/or removal for vulnerable surface water and GWUDI sources, the LT2ESWTR rules also require public water systems (PWSs) that have uncovered finished water storage facilities to either cover the storage facility or treat the water discharged from the reservoir before it enters the distribution system. The discharged water must be treated to achieve 4-log virus, 3-log *Giardia*, and 2-log *Cryptosporidium* inactivation and/or removal.

The LT2ESWTR applies to all public water utilities that use surface water or GWUDI, except those that purchase surface water and GWUDI. The LT2ESWTR has several requirements related to the use of UV disinfection for filtered and unfiltered water utilities and for uncovered reservoirs. The requirements for UV disinfection are summarized in this chapter, and this section discusses those that are specific to the UV dose required, validation testing, monitoring, reporting, and off-specification operation.

UV Disinfection Guidance Manual

Because UV disinfection was a relatively new technology in the US drinking water industry in the early 2000s, the USEPA agreed to publish UV disinfection guidelines as the UVDGM (USEPA 2006d). The purpose of the UVDGM is to facilitate planning, design, and operation of UV installations by familiarizing primacy agencies and utilities with important design, operation, and UV reactor validation issues. The UVDGM provides recommendations to assist utilities and engineering consultants in meeting the LT2ESWTR requirements and recommendations for states in implementing LT2ESWTR requirements. Elements of the UVDGM recommendations are incorporated throughout this book. The final UVDGM was published in November 2006 (USEPA 2006d).

UV Dose Requirements

UV dose requirements are established for up to 4 log of inactivation of *Cryptosporidium*, *Giardia*, and viruses, as shown in Table 4-2 (USEPA 2006a). These UV dose requirements are for postfilter applications of UV disinfection and for unfiltered water utilities that meet filtration avoidance criteria. The UV dose requirements represent the UV doses that are required for the various levels of inactivation of the specified microorganisms in controlled, bench-scale experiments (i.e., collimated beam tests). Adenovirus was used by USEPA to develop the virus UV dose requirements because of its resistance to UV disinfection. However, adenovirus is an exception, in that almost all viruses are inactivated to a level of at least 4 logs by UV doses <40 mJ cm^{-2}.

It should be noted that there are *validation factors* that need to be calculated and applied to the UV doses observed during full-scale validation testing to be compared to the required UV doses (as discussed in the next section).

Validation

The USEPA developed UV dose requirements for *Cryptosporidium*, *Giardia*, and viruses, previously described. The UV dose requirements account for uncertainty associated with the UV dose–response relationship of the microorganisms. However, in full-scale UV disinfection applications, other sources of uncertainty are introduced from the hydraulic effects of the UV installation, UV reactor equipment, and monitoring approach (e.g., UV sensors). Therefore, these UV dose requirements cannot be applied directly to a full-scale UV reactor because of the UV dose distributions that occur in the UV equipment and their associated uncertainties. Validation testing will verify reactor performance, and validation factors will be applied to account for these uncertainties in full-scale UV reactors. The validated UV dose, which includes these uncertainties, will be compared to the UV doses shown in Table 4-2 to determine the log credit achieved.

Table 4-2 UV dose requirements in mJ cm^{-2} (USEPA 2006a)

	UV Dose for Log Inactivation of							
Microorganism ↓	**0.5**	**1**	**1.5**	**2.0**	**2.5**	**3.0**	**3.5**	**4.0**
Giardia	1.6	2.5	3.9	5.8	8.5	12	15	22
Cryptosporidium	1.5	2.1	3.0	5.2	7.7	11	15	22
Viruses	39	58	79	100	121	143	163	186

The LT2ESWTR requires the following with respect to reactor validation (USEPA 2006a):

- Validation testing must determine a range of operating conditions for delivering the required UV dose that can be monitored by the UV equipment.
- At a minimum, operating conditions must include flow rate, UV intensity (as monitored by a UV sensor), and lamp status.
- Validated conditions determined by testing must account for UV absorbance of the water, lamp fouling and aging, measurement uncertainty of online UV sensors, UV dose distributions arising from the velocity profiles through the reactor, failure of UV lamps or other critical installation components, and inlet and outlet piping or channel configurations of the UV reactor.

Using these requirements as a basis, the UVDGM provides detailed guidance for several possible approaches to reactor validation. Chapter 6 describes the USEPA validation recommendations in more detail. Since UVDGM publication, additional factors and methods have been introduced, which will also be described in Chapter 6.

Off-Specification

For filtered and unfiltered water utilities using UV disinfection to meet the LT2ESWTR requirements, the required *Cryptosporidium* log inactivation by UV disinfection must be achieved in at least 95 percent of the water delivered to the public during each calendar month. If the required inactivation is not achieved, the UV reactors are operating off-specification. Off-specification events occur when the UV reactor operates outside of the validated conditions, the required validated UV dose or intensity setpoint is not achieved, or if the installed equipment is different from what was validated. Off-specification compliance is determined on a volume basis; the calculation methods are described in more detail in Chapter 10.

Monitoring and Reporting

The LT2ESWTR requires utilities to monitor their UV reactors to demonstrate that they are operating within the range of conditions that were validated for the required UV dose. At a minimum, utilities must monitor each reactor for flow rate, lamp outage, UV intensity as measured by a UV sensor, and any other parameters required by the governing agency. The UVT should also be measured when it is used in a UV dose control strategy. Utilities must check the calibration of UV sensors and recalibrate sensors in accordance with a protocol approved by the governing agency. The LT2ESWTR does not have a required monitoring frequency.

The LT2ESWTR requires utilities to report the following items:

- *Initial reporting*—Validation test results demonstrating operating conditions that achieve the UV dose required for the inactivation credit desired for compliance with the LT2ESWTR.

- *Routine reporting*—Volume of water entering the distribution system that was not treated by the UV reactors operating under validated conditions (i.e., off-specification) and the UV sensor calibration status on a monthly basis.

Chapter 10 describes the required and recommended monitoring and reporting in more detail.

GROUND WATER RULE

The GWR (USEPA 2006c) requires that groundwater utilities that are deemed vulnerable provide 4-log virus inactivation. The virus inactivation may be achieved with several possible technologies, including UV disinfection. The GWR states that UV disinfection can be used in a series configuration or in combination with another technology to provide 4-log virus inactivation. However, recent research studies show that validation of UV reactors for virus inactivation is possible; however, the GWR does not reflect this recent research. These studies are summarized as follows.

- Research (Linden et al. 2007) has shown that medium-pressure lamps achieved 4-log inactivation with UV doses lower than the LT2ESWTR UV doses for adenovirus, using a bench-scale test apparatus.
- Validation of UV reactors with adenovirus have been completed and showed high levels of virus inactivation with medium pressure lamps because adenovirus is susceptible to far-UV wavelengths (<230 nm) (Linden et al. 2009).
- New UV sensors have been designed for MP reactors that monitor wavelengths between 220 and 240 nm to monitor virus inactivation (UV supplier communications).
- New non-pathogenic challenge microorganisms are being researched that have closer dose-response to adenovirus (Sun, Zhe et al. 2020).
- Several medium pressure UV disinfection installations are receiving virus credit in the US (e.g., Atlantium, Calgon Carbon).
- New validation techniques have recently been published that could be used for validating virus inactivation (USEPA 2020).

REFERENCES

DIN 19294:1.2020. Devices for the Disinfection of Water Using Ultraviolet Radiation — Part 1: Devices Equipped with UV Low-Pressure Lamps — Requirements and Testing. Joint Austrian/German Standard. DIN. https://shop.austrian-standards.at/action/en/public/details/667583/OENORM_M_5873-1_2020_01_01.

Health Canada. 2020. Guidelines for Canadian Drinking Water Quality Summary Table [Online]. Available: https://www.canada.ca/en/health-canada/services/environmental-workplace-health/reports-publications/water-quality/guidelines-canadian-drinking-water-quality-summary-table.html.

Linden, K.; J. Thurston; R. Schaefer; J.P. Malley, Jr. 2007. Enhanced UV Inactivation of Adenoviruses Under Polychromatic UV Lamps. *Applied and Environmental Microbiology*, 73(23):7571–7574. doi:10.1128/AEM.01587-07.

Linden, K.G.; G.-A. Shin; J.-K. Lee; O.K. Scheible; C. Shen; P. Posy. 2009. Demonstrating 4-Log Adenovirus Inactivation in a Medium-Pressure Ultraviolet Disinfection Reactor. *Journal AWWA*, 101(4):90-97.

NSF/ANSI 55. 2020. Ultraviolet Microbiological Water Treatment Systems. NSF International. https://webstore.ansi.org/standards/nsf/nsfansi552020.

ÖNORM and DVGW. 2020. *UV Devices for Disinfection in Drinking Water Supply: Part I: Requirements on Properties, Function and Operation; Part 2: Testing of Properties, Function and Disinfection Effectiveness; Part 3: Ports and Sensors for the Radiometric Monitoring of UV Disinfection Devices—Requirements, Testing and Calibration*. DVGW (German Association for Gas and Water), Bonn, Germany. [Online] Available only in German: http://www.wvgw.de/index.php?id=451&submit_gesamtsuche=&tx_indexedsearch%5Bsword%5D=W+294.

Sun, Z.; Z. Zhi; E.R. Blatchley III. Using Algal Virus *Paramecium bursaria chlorella* virus (PBCV-1) as a Human Adenovirus Surrogate for the Validation of UV Treatment Systems. Proc. 2020 IUVA Americas Conference. March 8–11, 2020.

USEPA. 2020. Wright, H., T. Brooks and M. Heath. 2020. *Innovative Approaches for Validation of Ultraviolet Disinfection Reactors for Drinking Water Systems.* Report EPA/600/R20/094, USEPA, Washington, DC. https://cfpub.epa.gov/si/si_public_record_report.cfm?Lab=CESER&count=10000&dirEntryId=349759&searchall=&showcriteria=2&simplesearch=0&timstype=.

USEPA. 2006a. Long Term 2 Enhanced Surface Water Treatment Rule (LT2) [Online]. Available: https://www.epa.gov/dwreginfo/long-term-2-enhanced-surface-water-treatment-rule-documents.

USEPA. 2006b. Stage 2 Disinfectants and Disinfection Byproducts Rule (Stage 2 DBP Rule) [Online]. Available: https://www.epa.gov/dwreginfo/stage-1-and-stage-2-disinfectants-and-disinfection-byproducts-rules.

USEPA. 2006c. Ground Water Rule (GWR) [Online]. Available: https://www.epa.gov/dwreginfo/ground-water-rule.

USEPA. 2006d. *Ultraviolet Disinfection Guidance Manual for the Final Long Term 2 Enhanced Surface Water Treatment Rule* [Online]. Available: https://nepis.epa.gov/Exe/ZyPDF.cgi?Dockey=600006T3.txt.

CHAPTER 5

UV Disinfection Equipment

UV disinfection equipment must be designed so that all the microorganisms flowing through the reactor receive a UV dose (fluence) adequate to disinfect the water to an acceptable standard. This requires attention to the design of the UV reactor, the type and number of UV lamps, their placement in the UV reactor, any baffles that may be necessary to improve mixing, quartz-cleaning systems, and UV sensors.

TYPES OF AVAILABLE UV EQUIPMENT

The available UV equipment can be divided into two major types: *open-channel* and *closed-pipe* systems. The former are found primarily in wastewater and recycled water treatment facilities and the latter in drinking water treatment facilities.

Open-Channel Systems

In this system, the water to be treated flows by gravity through a channel with a rectangular cross-section with the water level maintained by a weir. The UV lamps are positioned either (or both) perpendicular and/or parallel to flow. In some open channel systems, particularly those that use medium-pressure UV lamps, the water is forced to flow through circular or octagonal channels, each of which contains a UV lamp and quartz sleeve. Figure 5-1 shows an example of this type of installation, where the UV lamps and quartz sleeves can be raised for cleaning and maintenance. Open-channel systems are not typically used in drinking water applications because UV disinfection is applied postfiltration, where the water should not be exposed to the environment.

Figure 5-1 An example of an open-channel UV disinfection system. This unit consists of 64 medium-pressure UV lamps treating wastewater at 20 mgd. The UV lamps and their quartz sleeves can be swung up for lamp replacement and maintenance. This unit is located in London, Ont., Canada. (Photo courtesy of Trojan Technologies)

Closed-Pipe Systems

In this system, the UV disinfection system consists of units with a circular or rectangular cross-section that are inserted into the water flow (i.e., into a pipe) in a water treatment facility. The units contain UV lamps, UV sensors, quartz sleeves around the lamps, and usually a quartz-cleaning mechanism. There are many types of closed-pipe UV reactor, but most can be classified into the following types:

Type 1: *One-lamp annular reactor* (Figure 5-2a). This is the simplest type and consists of a cylindrical outer wall with a single lamp and quartz sleeve along the axis of the cylinder. The water flows from one end, through the annular space between the quartz sleeve and the outer wall, and out the other end. This type of reactor is commonly found in low flow rate (1–10 gpm) point-of-use (POU) systems, such as those for disinfecting the drinking water in a home, boat, or cottage. Most UV reactors of this type use low-pressure UV lamps.

Type 2: *Multiple lamp reactor with lamps parallel to flow* (Figure 5-2b). This is a cylindrical cross-section UV reactor, similar to Type 1, but several lamps with quartz sleeves are located parallel to the axis of the cylinder. This type of reactor is designed for a higher flow rate (10–500 gpm) than Type 1 and would be suitable for an apartment building or very small community. Most UV reactors of this type also use low-pressure or low-pressure, high-output UV lamps.

Type 3: *Multiple lamp reactor with lamps perpendicular to flow* (Figure 5-2c). These lamps with their quartz sleeves traverse the axis of the reactor and are perpendicular to the water flow. The reactor cross-section can be circular, square, or rectangular. For this type of reactor, often medium-pressure lamps are used, although some use low-pressure or low-pressure, high-output lamps. The lamps can be oriented horizontally, vertically, or at some angle in between. This type of reactor can handle much higher flow rates (500 gpm to 50 MGD) and is often found in medium to large water treatment plants.

UV LAMPS

The most important components of UV disinfection equipment are the UV lamps. These lamps must be able to generate UV radiation in the germicidal region (200–300 nm) with acceptable efficiency and lifetime.

A UV lamp converts electrical energy partially into light energy with the rest dissipated as heat. The lamp efficiency is usually defined as the ratio of the total UV power output in the germicidal region to that of the external electrical power input.*

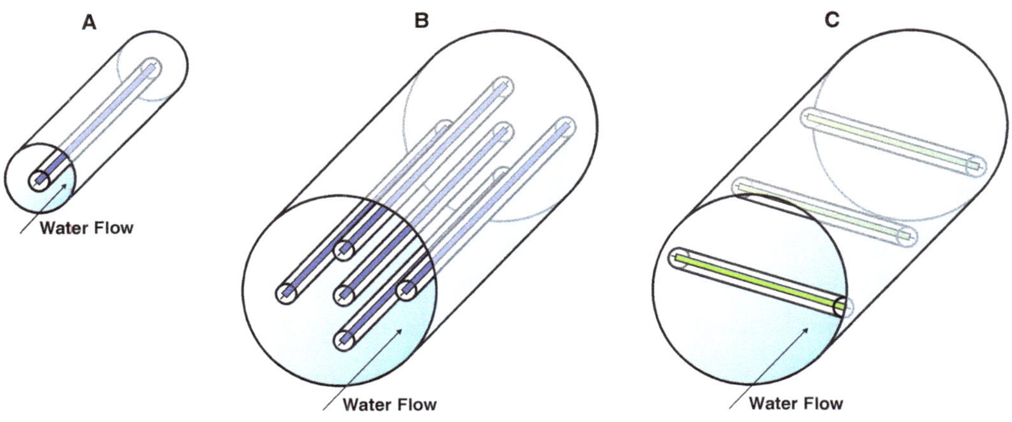

Figure 5-2 Closed-pipe UV reactor types: simple annular one-lamp reactor (A); multiple lamp reactor with lamps parallel to flow (B); multiple lamp reactor with lamps perpendicular to flow (C).

* Some manufacturers use the *electrical power to the lamp* as the basis for efficiency specifications. This definition neglects the power losses in the power supply.

Table 5-1 Classification of UV lamps

Lamp Type	Characteristics	Wavelength Range / nm	Example
Temperature radiator lamps	Hot filament (e.g., tungsten) in an inert gas	350–800	Incandescent light bulb
Gas discharge lamps	Electrodes at each end of a long cylindrical lamp containing a metal vapor (e.g., mercury)	150–800	Fluorescent light bulb; low pressure and medium pressure mercury lamps
Light-emitting diodes (LEDs)	Solid-state semiconductors that emit light when current is passed through	250–700	Indicator lights on electronic equipment
Excilamps	Electric discharge in a lamp containing an inert gas (e.g., Xe) or a halogen (e.g., Cl_2)	Relatively monochromatic in the range 170–300 nm	New experimental type of lamp

Lamp Types

There are several types of lamps that emit in the UV and visible region (see Table 5-1).

The most important lamps for UV disinfection are gas discharge lamps. These usually have two electrodes, one at each end of a tube containing a gas, which may contain metal atoms (e.g., mercury) or a rare gas (e.g., xenon). In such lamps, light emission arises from atoms that have been excited to higher energy states by high-energy electrons emitted from the electrodes. There are several common types of gas discharge lamps.

The vast majority of lamps used in UV disinfection systems are mercury vapor lamps. In these lamps, electric current that flows through the ionized hot plasma causes the mercury atoms to be raised to higher energy or excited states. When these excited states return to their ground state, UV radiation is emitted with a wavelength inversely proportional to the energy difference between the excited state and the ground state.

The lifetime of the excited mercury atoms is limited and in microseconds (μs) or milliseconds, they decay to lower energy levels with the excess energy emitted as light or heat. The wavelength of the light is determined by the difference in the energy levels (see Eq. 2-1b). The relative intensity of the various transitions varies considerably.

There are four common types of mercury discharge lamps: low-pressure (LP); low- pressure, high-output (LPHO); LPHO-amalgam; and medium-pressure (MP) lamps.

Low-Pressure Mercury Lamps

LP lamps operate at temperatures of 20–40°C, where the mercury vapor pressure is about 0.1–10 Pa (approx. 0.001–0.1 torr). These lamps emit only in narrow lines with virtually no emission between the lines (see Figure 5-3). Figure 5-4 shows the wavelengths and relative emittances for the lines of a low-pressure mercury vapor lamp. The strongest lines in the UV are those at 253.7 nm (82 percent of the total emission) and at 184.9 nm (6.6 percent of the total emission).

The LP lamp, containing only a few milligrams of mercury, is very common. The principal emission is at 253.7 nm (see Figure 5-4), which effectively inactivates microbial pathogens; therefore, this lamp is called a *germicidal* lamp. The output power is rather low; for example, a 4 ft (122 cm) LP lamp is rated at approximately 40 W. The inner surface of the lamp envelope can be coated with various types of *phosphors*, which absorb the 253.7 nm radiation and emit at longer wavelengths. This is the basis of the very popular *fluorescent lamps.*

The light output of an LP lamp depends sharply on the bulb temperature. The optimum temperature depends somewhat on lamp design but is usually about 40°C. Figure 5-5 shows the temperature dependence of the output of a low-pressure lamp in open air.

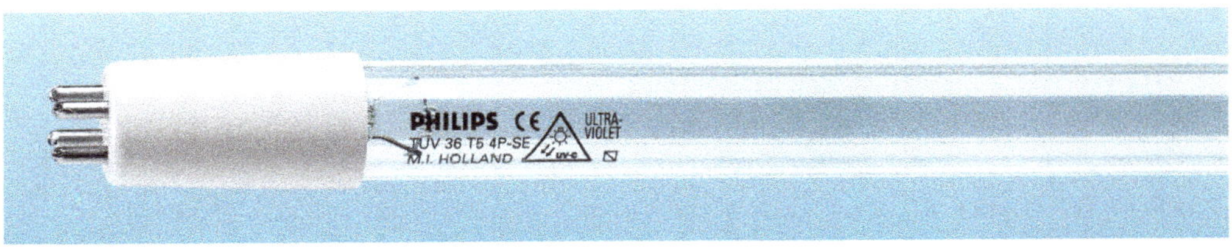

Figure 5-3 Picture of a standard LP mercury lamp (courtesy of Philips Lighting)

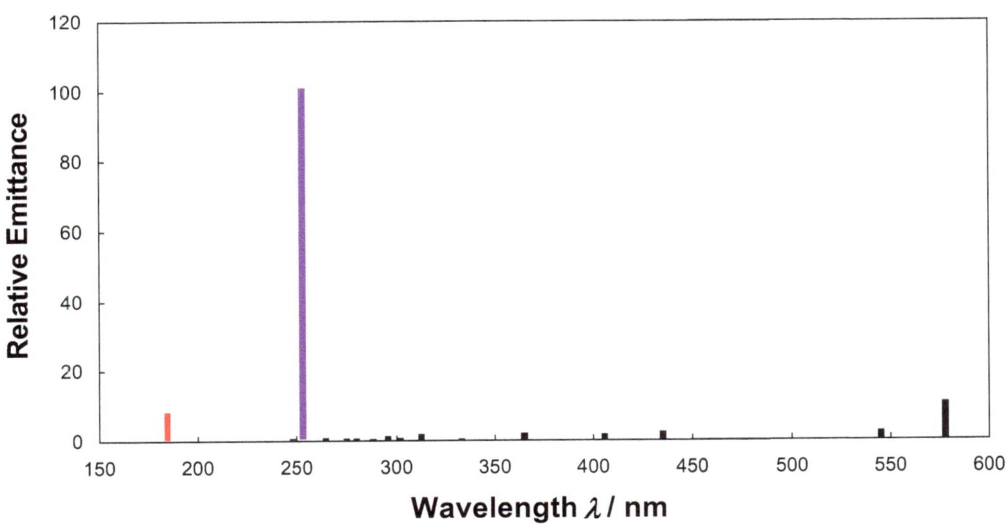

Figure 5-4 Relative emittance of the various lines from an LP mercury lamp

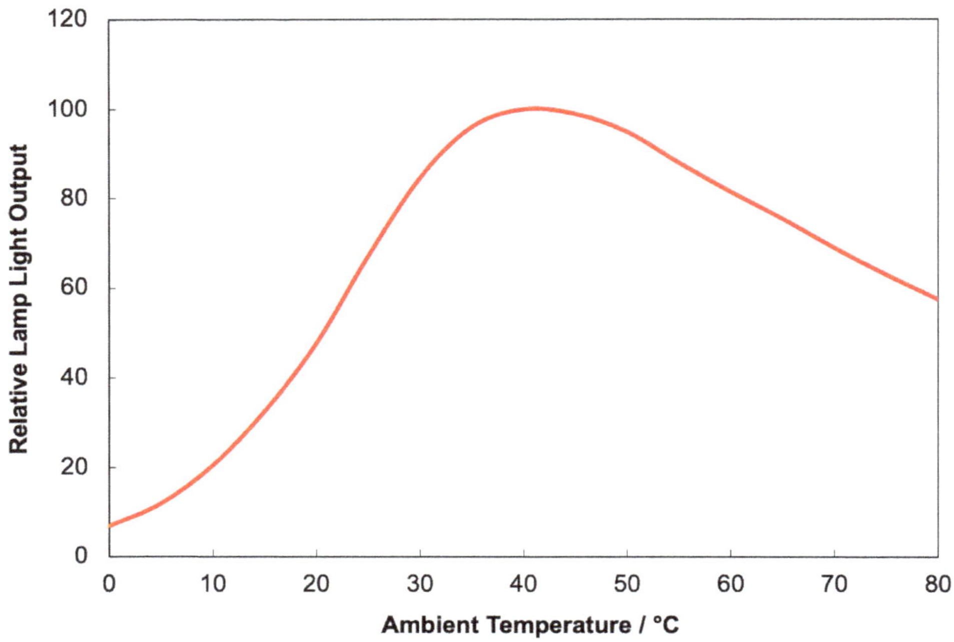

Figure 5-5 Relative lamp UV radiation output of an LP lamp in open air versus the ambient temperature (courtesy of Philips Lighting)

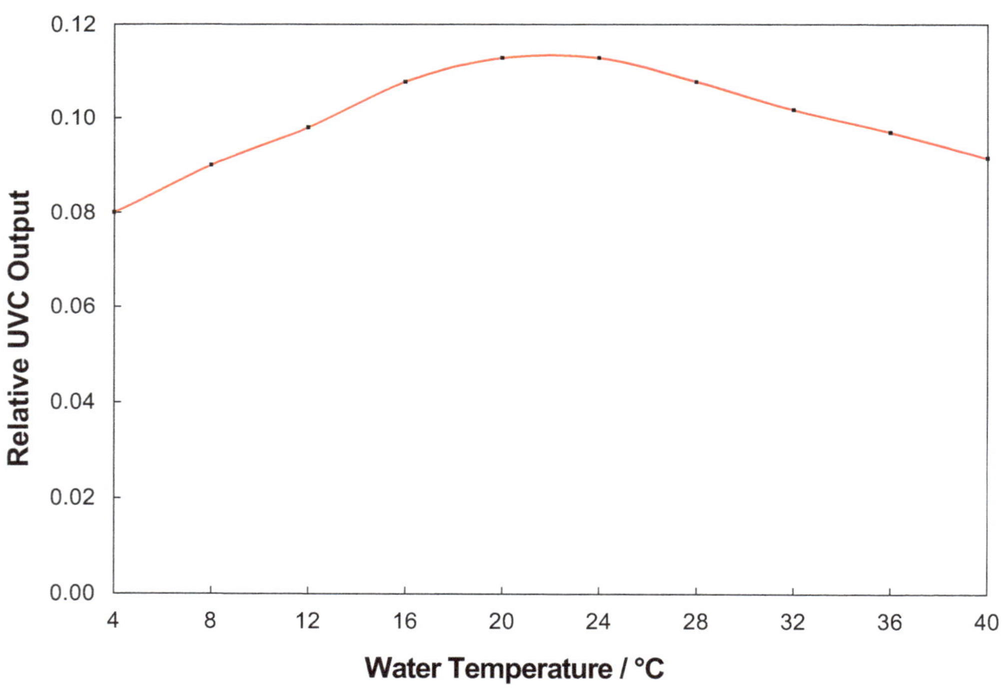

Figure 5-6 Relative UVC output from an LP UV lamp in a quartz sleeve (22 mm OD) as a function of the water temperature. The UV lamp was operated at a current of 425 mA.

The temperature around the lamp in a UV reactor for most water disinfection applications can be controlled to a certain extent by the diameter of the quartz sleeve separating the lamp from the water. Because of the limited heat transfer between the lamp and the water, the UV output becomes less sensitive to the water temperature. An example is shown in Figure 5-6. However, there are start-up problems in northern climates, where water temperatures can be as low as 0.5°C in the winter.

Low-Pressure, High-Output Mercury Lamps and Amalgam Lamps

Since about 1998, two types of modified low-pressure lamps have been introduced. They have the same monochromatic output at 253.7 nm as an LP lamp but have from 1.5 to 3 times the output power per unit length.

The first type is called an LPHO lamp. It consists of a standard LP lamp but with reinforced filaments standing out 40–60 mm from the end of the lamp. This configuration allows the lamp to maintain a current 1.5–2 times that of a standard LP lamp without heating the electrodes too much. If the electrodes get too hot, the mercury vapor pressure rises and the overall efficiency drops.

A more recent type of enhanced low-pressure lamp is the amalgam lamp† (see Figure 5-7). These lamps have heavy electrodes and contain no "free" mercury. A solid amalgam (a compound of mercury with another element, such as indium or gallium) spot is placed on the inner wall of the lamp. This spot serves to control the mercury vapor pressure, even though a higher current is flowing. Amalgam lamps can have a UVC output 2–3 times that of a conventional LP lamp.

Because LPHO and amalgam lamps operate at temperatures around 100°C, their output is less sensitive to the water temperature than is the case for LP lamps.

Vacuum UV Lamps

If an LP or LPHO lamp is manufactured with an envelope made from ultrapure quartz, the lamp can transmit the emission line at 184.9 nm, which is in the vacuum UV region. This emission is strongly absorbed by O_2 in the air (to generate ozone) and by liquid water (generating •OH radicals).

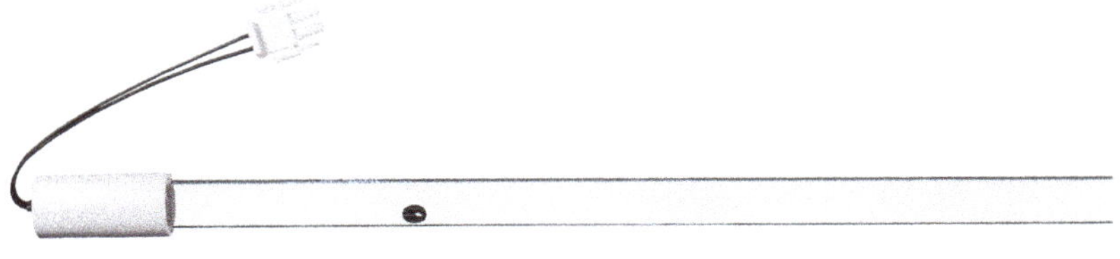

Figure 5-7 Picture of an amalgam lamp. The spot on the inner surface of the lamp is the solid mercury amalgam that controls the mercury vapor pressure. (Courtesy of Philips Lighting)

† Unfortunately, some manufacturers often also call this type of lamp *LPHO lamp*..

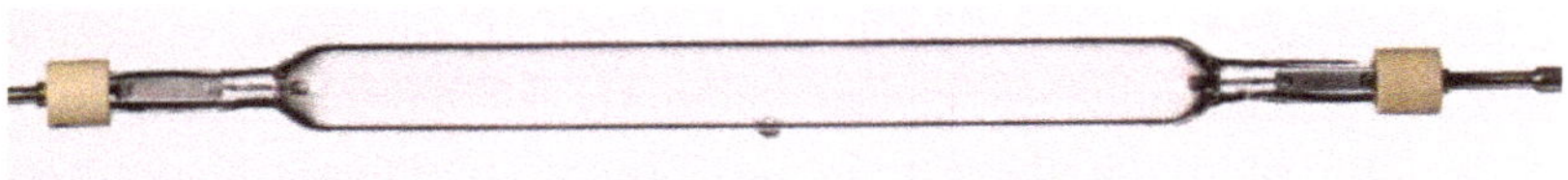

Figure 5-8 Picture of a 2.5 kW MP UV lamp (Courtesy of Heraeus Noblelight GmbH)

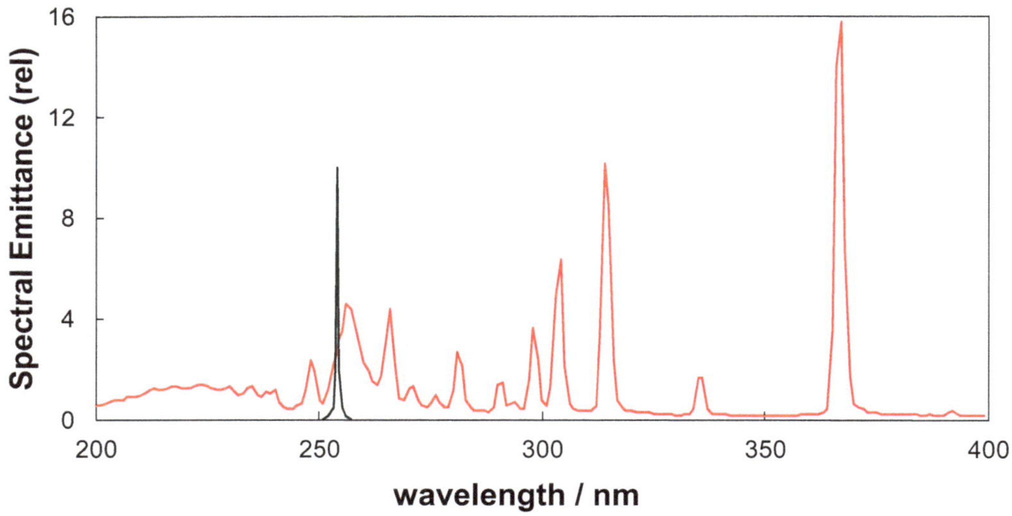

Figure 5-9 Relative spectral emittance from LP and MP lamps (Bolton, personal communication)

Medium-Pressure Mercury Lamps

The emission lines of a mercury lamp are only sharp when the pressure of the gas is low (<100 Pa). If the pressure is increased, the lamp can carry much more power, but the increased collision rate from other gas molecules causes the emission lines to broaden. For the same length of lamp (about 120 cm), an MP lamp (pressure about 100 kPa or about 1 atm) can carry up to 30,000 W (see Figure 5-8). MP lamps are also commonly used in commercial systems using UV radiation. Figure 5-9 shows a comparison of the emission of LP and MP lamps in the UV region. MP lamps operate very hot; thus, they are not overly sensitive to the ambient temperature.

Flash Lamps

Flash lamps are similar to continuous wave (CW) arc lamps, in that they consist of a cylindrical quartz tube with electrodes at each end and are filled with a gas (e.g., xenon). A power supply *fires* the lamps by discharging a large amount of electrical energy (100–1,000 J) in a very short period of time (several microseconds) by applying a very high voltage (10–30 kV). The resulting plasma reaches temperatures of 8,000–25,000 K, and the emission is essentially that of a blackbody.‡

‡ All objects at temperatures above 0 K emit radiation, called *blackbody* emission. The peak wavelength shifts to shorter wavelengths as the temperature increases.

Table 5-2 Emission wavelengths for some common excilamps (Oppenländer 1994)

Excimer	Wavelength λ / nm	Excimer	Wavelength λ / nm
Ar_2^*	172	Cl_2^*	259
Kr_2^*	146	$XeBr^*$	283
Xe_2^*	172	Br^*	289
$KrBr^*$	207	$XeCl^*$	308
$KrCl^*$	222	I_2^*	342
XeI^*	253	XeF^*	354

* The entity is in an electronically excited state.

In some experimental flash lamp systems, the lamp is flashed about 30 times per second. At present, the germicidal efficiency and lifetimes are low, and very few commercial systems use flash lamps.

Excilamps

Excilamps (also called excimer lamps) are unique in that they emit in a narrow band of wavelengths. They involve the formation of excimers, which are molecular dimers that are stable only in the excited state and dissociate when decaying to the ground state. Table 5-2 gives the wavelengths of some of the common excimer lamps. Oppenländer (1994) and Sosnin et al. (2006) describe these lamps and their applications. Excilamps are now only in the research stage, but if the efficiencies and lifetimes improve significantly, they could become viable commercially.

UV-LEDs

Recently, Light-Emitting Diodes emitting in the UV region (UV-LEDs) have become available. These are semiconductor devices that emit UV in a narrow bandwidth (~10 nm) centered at wavelengths from 250–400 nm. Because the optical output of UV-LEDs drops off at higher temperatures, UV-LEDs usually have to be attached to a heat sink or be cooled by water or air flow. Generally, the UV output efficiency is low (<10 percent) and decreases as the center wavelength decreases.

UV-LEDs have some advantages:

1. They are small and can be located almost anywhere in the UV reactor.
2. The center wavelength can be chosen to match the peak germicidal factor for a specific microorganism.
3. LEDs use electricity to generate UV light and are mercury-free.

At present, UV-LEDs in the germicidal wavelength range are not very efficient (<2 percent). However, as with any semiconductor device, the UV output efficiency and lifetime are expected to increase significantly as time passes.

UV reactors employing UV-LEDs are now becoming available, often with interesting design features. The size and flow rate are usually small, but there is one case of a full-scale drinking water UV reactor (250 m^3/h or 1,100 gpm) employing 1000 UV-LEDs (Autin and Bolton 2019). Please note LEDs can be developed at specific wavelengths, which LADWP did as part of its nitrification prevention research (LADWP 2021).

Comparison of Lamp Types

Table 5-3 gives a comparison of the characteristics of the principal lamp types that are used in UV disinfection reactors. It should be noted that the LP, LPHO, and amalgam lamps have the highest germicidal efficiency. This means that LP or LPHO lamps are the choice for very small flow rates (<100 gpm). LPHO and amalgam lamps are often used for mid-size to high flow rates (0.1–10 mgd and larger). MP lamps are often used for very high flow rates (10 mgd and larger) because the fewer lamps required means much lower maintenance costs, although some large flow rate applications use LPHO UV lamps.

An important factor to consider in the choice of lamp type is the germicidal lamp efficiency. Clearly, LP and LPHO lamps have a marked advantage because their efficiencies are approximately twice that of MP lamps. However, the overall

Table 5-3 Comparison of the characteristics of UV lamps used for UV disinfection of drinking water

Characteristic	Low Pressure (LP)	LPHO*	Medium Pressure	Flash (Pulsed)	UV-LEDs
Emission	Virtually mono-chromatic (253.7 nm)	Virtually mono-chromatic (253.7 nm)	Poly-chromatic (185–600 nm)	Poly-chromatic (185–800 nm)	Narrow band of wave-lengths
Mercury vapor pressure (Pa)	0.1–10 Pa	0.1–10 Pa	50–300 kPa	NA	NA
Mass of mercury for a 1.2 m long lamp	~30 mg	~30–75 mg	2–4 g	NA	NA
Operating bulb temperature	30–50ºC	60–100ºC	600–900ºC	Plasma temp. 8,000–25,000 K	Close to ambient temp.
Arc length (cm)	15–200	15–200	10–200	10–100	NA
Lifetime	8,000–12,000 h	7,000–10,000 h	5,000–8,000 h	>10^8 pulses	<3000 h
Germicidal efficiency†	35–40%	30–35%	12–16%	Unknown	<10%
Power density (input power in watts per cm)	0.2–0.4	0.6–1.2	125–200	NA	NA

*LPHO lamps are either amalgam lamps or lamps with special ends that allow a cold spot at the ends of the lamp.

† Weighted to the relative spectrum of DNA.

life cycle cost should be considered, including maintenance, because the benefit of a higher efficiency may be offset by lower capital and maintenance costs. In addition, there are often site-specific costs that will influence the choice of lamp type.

UV SENSORS

The UV sensor is an important component of a UV reactor because each UV sensor monitors the irradiance (or intensity) at specific locations in the reactor. During the validation of a UV reactor, the UV sensor reading is correlated with the UV dose (fluence) delivered by the reactor and thus becomes an important part of the online monitoring system.

Most UV sensors utilize a light-sensitive semiconductor (e.g., SiC) that is *solar blind*. This means that the sensor is not activated by solar radiation, which has wavelengths >300 nm. Often a filter is applied so that the sensor is active primarily in the germicidal range of about 200–300 nm. Figure 5-10 shows pictures of UV sensors.

UV sensors continuously monitor the irradiance at a specific position in the UV reactor; however, the sensor reading is affected by several factors: (1) the UV output of the lamp; (2) the transmittance of the quartz sleeve (fouling of this sleeve can drastically decrease the UV radiation that enters the reactor); (3) the transmittance of the water (an increase in absorbing substances in the water will decrease the sensor reading); and (4) the transmittance of the sensor window (again, fouling of this surface can drastically reduce the sensor reading). Proper maintenance of the UV reactor should minimize the effects of factors 2 and 4 so that factors 1 and 3 are the most significant.

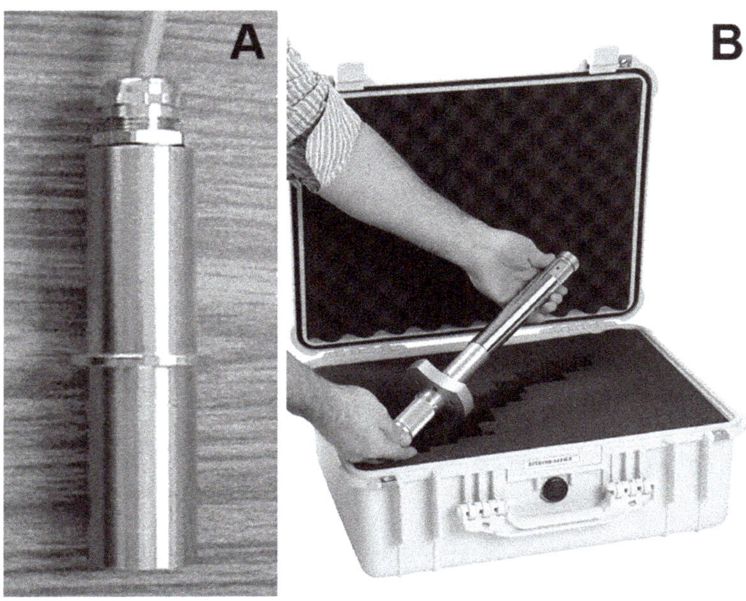

Figure 5-10 Pictures of UV sensors (A: courtesy of Calgon Carbon Corporation; B: courtesy of Trojan Technologies)

UV sensors must be checked against a reference sensor every month, and the reference sensor must be calibrated periodically (recommended in the UVDGM to be performed once per year). This is performed using a reference sensor that is identical to the duty sensor, except that it has been calibrated against national standards, such as those maintained by the National Institute for Standards and Technology (NIST). This assures that the UV sensors are reading irradiances accurately.

SLEEVES

A UV lamp must be separated by an air space from the flowing water because a UV lamp needs to operate at a higher temperature than the water. This separation is accomplished by placing the lamp inside a sleeve, which is usually made of quartz because quartz is one of the few materials that transmit UV in the 200–300 nm region. However, some sleeves use a high-purity form of polyfloro-ethylene (Teflon), but the percent transmittance of Teflon sleeves is less than that of quartz.

Cleaning of Sleeves

Most mineral salts (e.g., calcium, magnesium, and iron salts) in drinking water have a solubility that decreases with increasing temperature. In addition, photochemical reactions can cause mineral deposition. Thus the quartz sleeve, which is close to a hot UV lamp, has a higher temperature than the surrounding water, providing a place where these salts may precipitate. This fouls the quartz sleeve and decreases the UV transmittance. Therefore, it is usually necessary to design UV reactors with an effective quartz cleaning mechanism. In small UV reactors, it may be sufficient to introduce a maintenance schedule that includes periodic removal (e.g., 2–3 times per month) of the quartz sleeve that is manually cleaned with an acid solution. In larger UV reactors, it is necessary to have an on-line quartz cleaning system. This can either be mechanical (e.g., utilizing a stainless-steel brush) or chemical (e.g., using a sealed acid solution) or a combination of both methods. Periodically, the cleaning system should move across the quartz sleeve to clean off any deposits that may have formed. One manufacturer uses sonic cleaning. Some larger UV reactors can also have off-line chemical cleaning systems that spray an acid solution into the reactor. These systems require the reactor to be taken off-line, but they do not require the lamp sleeves to be removed for cleaning. Figure 5-11 shows pictures of each of the quartz cleaning systems.

ANCILLARY COMPONENTS

There are several other components that are essential to the operation of a UV reactor.

1. *Lamp power supply*. This system is usually external from the UV reactor itself and provides the proper voltage and current to operate each

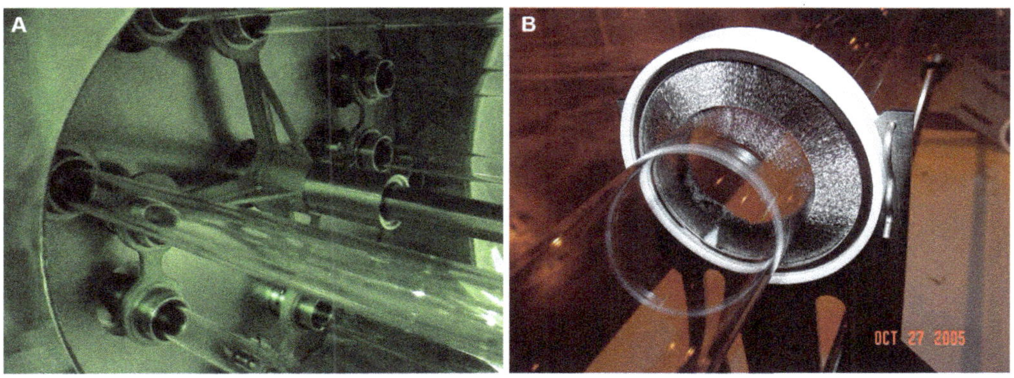

Figure 5-11 Pictures of quartz cleaning systems: chemical cleaning system (A: courtesy of Trojan Technologies); mechanical system (B: courtesy of Calgon Carbon Corporation)

UV lamp. Most UV lamps require a ballast to provide the starting voltage and to stabilize the current during operation. This is usually done passively using an inductor; however, there now exist electronic ballasts that actively control the operation of the UV lamps and may operate at a much higher frequency than the line frequency. The power supply exhibits a certain efficiency (about 92–95 percent for inductive ballast power supplies and 95–98 percent for electronic power supplies), defined as the ratio of the power across the lamp to the power drawn from the wall.

2. *Control of UV reactors.* Several factors need to be known in the control center, so operators can be assured that the UV reactor is operating to achieve the disinfection target. The most important readings are UV sensor readings and the flow rate; however, it is also desirable to monitor the UV transmittance (UVT at 254 nm for a 1-cm path length) using an online transmittance analyzer if the calculated dose control approach (see Chapter 6) is used. Some manufacturers have developed an algorithm that can be programmed into a programmable logic controller (PLC) that takes the inputs of UV sensor readings, flow rate, and UVT and provides an estimate of the UV dose, lamp status, and alarm output in the control center.

FULL-SCALE OPERATING UV DISINFECTION SYSTEMS

Since 1998, there has been a growing trend for medium to large water treatment plants to install UV disinfection systems, not only to comply with USEPA regulations (e.g., the LT2ESWTR), but also to provide a second disinfection barrier, even when regulations do not require it.

Figures 5-12 through 5-15 illustrate a variety of medium to large UV disinfection installations.

Figure 5-12 Trojan Technologies UV disinfection system installed in Victoria, B.C., Canada. Flow rate 145 mgd (21,292 m^3/h). Each reactor contains 8 × 10 kW MP UV lamps. There are 15 reactors in the system. (Picture courtesy of Trojan Technologies)

Figure 5-13 Wedeco K-series UV disinfection system installed in Clayton County, Ga. Flow rate 12 mgd (1,893 m^3/h). Each reactor contains 90 × 150 W LPHO UV lamps. There are three reactors in the system. (Picture courtesy of ITT Wedeco)

Figure 5-14 Calgon Carbon UV disinfection system installed in Edmonton, Alta., Canada. Flow rate is 95 mgd (14,982 m^3/h). Each reactor contains six MP 20 kW UV lamps. There are three reactors in the system. (Picture courtesy of Calgon Carbon Corporation)

Figure 5-15 Full-scale installation of UV reactors (6) at the Cumwhinton Water Treatment Works in Carlyle, UK. Each UV reactor has 1,000 UV-LEDs emitting at 275 nm. The rated flow rate is 28 MLD (106 MGD). (Courtesy of Typhon Treatment Systems, Inc.)

REFERENCES

Autin, O. and J.R. Bolton. 2019. Validation of a reactor containing UV-LEDs for the disinfection of municipal drinking water. *UV Solutions*, Q3:17–23.

Los Angeles Department of Water and Power (LADWP), 2021. Personal communication.

Oppenländer, T. 1994. Novel Incoherent Excimer UV Irradiation Units for the Application in Photochemistry, Photobiology/Medicine and for Waste Water Treatment. European Photochemistry Association (EPA), Zürich, Switzerland. *EPA Newsletter*, March: 2–8.

Sosnin, E.A.; T. Oppenländer; V.F. Tarasenko. 2006. Applications of Capacitive and Barrier Discharge Excilamps in Photoscience. *Journal of Photochemistry and Photobiology C*, 7:145–163.

CHAPTER 6

Validation of UV Reactor Performance

UV reactors have variations in performance, depending on the water quality, flow rate, lamp performance and power level, and UV reactor hydraulics. The UV dose (fluence) in a reactor cannot be monitored directly, unlike chemical disinfectants where a disinfectant residual can be measured. There are two techniques widely used to assess UV reactor performance—biodosimetry and modeling. Typically full-scale biodosimetry testing (i.e., validation testing) is completed by the UV reactor manufacturer. These techniques and appropriate regulatory requirements are described in this chapter. Other emerging techniques, such as dyed microspheres, are currently being developed for reactor validation; however, these techniques are not described in this chapter since they were still in the research phase and not in wide-spread use.

REGULATORY REQUIREMENTS

The LT2ESWTR (see Chapter 4) requires that UV reactors are validated using biodosimetry with a challenge microorganism on a full-scale UV reactor. Alternative validation approaches, including modeling and dyed microspheres, are not currently accepted by the USEPA as a method of verifying performance. However, the USEPA recognizes these approaches may become more robust in the future and gives flexibility to the governing agency to approve alternative validation methods.

The UV equipment tested during validation must conform uniformly to the UV equipment being used by the water utility. The UV dose (fluence)–response characteristics of the challenge microorganism must be determined using a bench-scale, collimated beam test with an LP lamp (described in more detail in Chapter 3).

Validation testing is required to determine a set of operating conditions that a utility can monitor to ensure that the UV dose (fluence) required for a given level of pathogen log inactivation credit is achieved. The utility must then monitor the reactor to demonstrate that it is operating within the range of validated conditions.

At a minimum, the following UV reactor operating conditions must be determined during validation testing:

- UV intensity (irradiance);
- flow rate; and
- lamp status.

The UV equipment, water quality, and UV facility configuration will affect the performance of the UV reactor and must be accounted for in the UV reactor operating conditions. Specifically, the USEPA requires that the following factors be taken into consideration when developing the UV reactor operating conditions during validation testing:

- lamp aging;
- lamp sleeve fouling;
- UVT of the water;
- inlet and outlet piping or channel configurations of the UV reactor;
- UV dose (fluence) distributions resulting from the velocity profile of the water passing through the UV reactor;
- failure of UV lamps or other critical system components; and
- measurement uncertainty of on-line sensors (both intensity [irradiance] and UVT).

In addition to these requirements, the USEPA has developed a detailed, recommended validation protocol that ensures these requirements are met, as summarized in this chapter.

The USEPA allows UV reactors previously validated under certain existing protocols to receive log inactivation credit, which includes the Austrian standards ÖNORM M 5873-1 and M 5873-2 (ÖNORM 2001, 2003) and the German guideline DVGW W294 (DVGW 2006). The validation must provide data on UV dose (fluence) delivery and monitoring for a documented UV reactor, and the analysis of those data must be used to relate the measured performance to a required pathogen log inactivation credit. UV reactors certified by DVGW and ÖNORM for a *B. subtilis* RED of 40 mJ cm^{-2} (400 J m^{-2}) (the RED concept is described later in this chapter) can be granted 3-log *Cryptosporidium* and *Giardia* inactivation credit. Please note there is a new ÖNORM/DVGW standard (DIN 2020) that supersedes these. However, the USEPA recommends that validations that follow the National Water Research Institute (NWRI)/Water Resource Foundation (formerly AwwaRF) guidelines and NSF Standard 55 be evaluated on a case-by-case basis. The new ÖNORM/DBGW 2020 standard (DIN 2020) would also need to be reviewed. The UVDGM provides guidance on how to assess validation data using these protocols (USEPA 2006).

BIODOSIMETRY PROCESS

Biodosimetry is a biological method of assessing performance of a UV reactor. Although the target microorganisms for water treatment plants (WTPs) are typically *Giardia*, *Cryptosporidium*, plus viruses and bacteria, it is difficult, costly, and hazardous to evaluate UV reactors with these microorganisms. Therefore, a nonpathogenic challenge microorganism is typically used in the biodosimetry process. Currently, the most common challenge organisms are MS-2 bacteriophage, *B. subtilis*, and T1 viruses. Biodosimetry uses full-scale testing and bench-scale analysis with a challenge microorganism, which are completed in tandem, to verify UV reactor performance. In the full-scale tests, the challenge microorganism is injected upstream from the UV reactor and samples are taken of the influent and effluent water so that the log inactivation can be determined for various conditions of water quality, lamp power, and water flow.

The bench-scale tests are completed separately from the full-scale testing. The purpose of these tests is to determine the UV dose (fluence)–response of the challenge microorganism in the same spiked water that passed through the UV reactor. Bench-scale tests are typically completed using a collimated beam apparatus with an LP lamp to determine the UV dose (fluence)–response curve of the challenge microorganism (i.e., log inactivation versus UV dose).

The log reduction (inactivation) observed in the full-scale tests is correlated to a RED from the bench-scale collimated beam test that is specific to the challenge microorganism. Full-scale tests need to be completed for a matrix of water quality and flows, and the RED is calculated at each condition. The RED is then adjusted for uncertainty incurred during the biodosimetry process to determine the validated UV dose (fluence). An overview of the biodosimetry process is shown in Figure 6-1.

OVERVIEW OF USEPA-RECOMMENDED APPROACH*

The USEPA requires that biodosimetry be used to verify reactor performance. In addition, the UVDGM recommends a specific approach for determining what log inactivation credit the UV reactor will receive based on the validation data.

There are factors that create uncertainty in translating the RED measured during a validation test to the UV dose (fluence) delivered at the utility during routine operation, including hydraulics through the reactor, experimental variability, and UV sensor measurement errors. To account for this uncertainty, the USEPA recommends that a validation factor (VF) be applied to the measured, challenge microorganism RED value to determine the validated UV dose (fluence). RED values are specific to the challenge microorganism's inactivation kinetics and action spectra, the lamp spectral output, and the UV reactor's UV dose distribution. Therefore, REDs for different challenge microorganisms cannot be directly compared even though they have the same dimensions (i.e., mJ cm^{-2}).

---

* This section is summarized and adapted from the UVDGM (USEPA 2006).

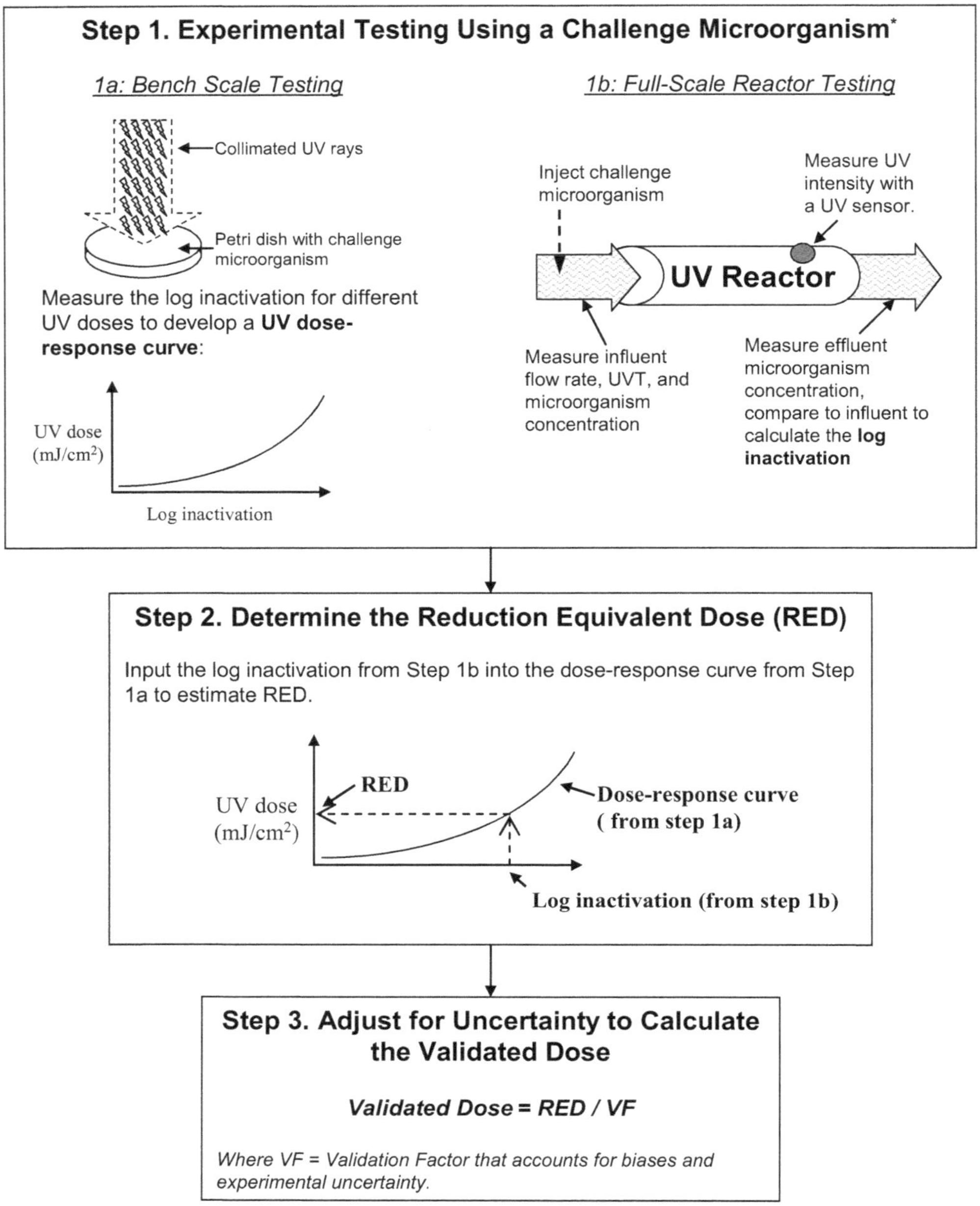

Figure 6-1 Biodosimetry process (USEPA 2006)

$$\text{Validated UV dose} = \frac{\text{RED}}{\text{VF}} \qquad \text{(Eq 6-1a)}$$

$$\text{VF} = B_{\text{RED}} \times (1 + U_{\text{VAL}}/100) \times \text{CF}_{\text{AS}} \qquad \text{(Eq 6-1b)}$$

where:

VF = validation factor

B_{RED} = RED bias factor

U_{VAL} = uncertainty in validation (expressed as a percent)

CF_{AS} = action spectra correction factor to account for differences in the challenge and target microorganism when using MP lamp reactors [Please note this was not included in the UVDGM and was recommended to be accounted based on new information (Linden et al. 2015).]

The UVDGM provides the detailed methods for determining the first three factors but not the action spectra correction factor. However, one must realize that the USEPA specifically addresses how to validate for two UV reactor control strategies (USEPA 2006).

- *The UV Intensity (Irradiance) Setpoint Approach.* Eq 6-1 produces a single validated UV dose (fluence) for a given UV intensity (irradiance) set-point corresponding to the minimum RED. In this approach, the control system sets off an alarm (indicating noncompliance) if any sensor reading (UV intensity [irradiance]) drops below a critical value (the setpoint). With this approach, the UV equipment is usually *overdosing* because the sensor reading will usually be significantly higher than the setpoint value. It is possible to also develop multiple setpoints to reduce the energy use.
- *The Calculated UV Dose Approach.* The validated UV dose (fluence) varies depending on operational parameters, such as values of the UVT, sensor reading, and flow rates. Usually, an equation is derived from the validation data that relates the RED to the operational parameters. The RED is then divided by the VF (Eq 6-1) to derive the validated UV dose (fluence). The advantage of this approach is that the WTP can save power by operating closer to the required UV dose (fluence).

The *RED bias* is a correction that accounts for the difference between the UV sensitivity of the target pathogen and the UV sensitivity of the challenge microorganism. This bias occurs as a result of the differences between the UV inactivation kinetics of the pathogen (e.g., *Cryptosporidium*) and the challenge microorganism (e.g., MS2) and the UV reactor UV dose distribution (Figure 6-2).

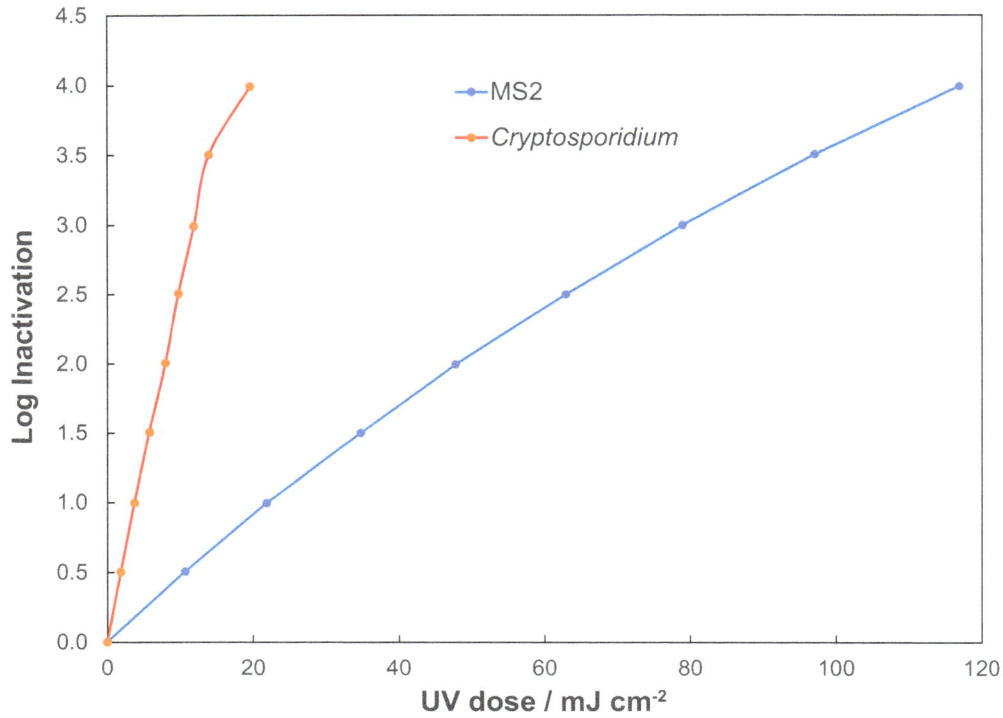

Figure 6-2 UV dose–response curves for male-specific-2 bacteriophage and *Cryptosporidium* (USEPA 2006)

Table 6-1 Calculated RED based on ideal UV dose distribution and UV dose–response curves

UV Dose (mJ cm^{-2})	Occurrence Probability (A)	*Cryptosporidium* Sensitivity† (mJ cm^{-2} Per Log of Inactivation)	*Cryptosporidium* Log Inactivation (B)	MS2 Sensitivity† (mJ cm^{-2}/ Per Log of Inactivation)	MS2 Log Inactivation (C)
5	0%	2.8	0	25.0	0
7.5	0%	3.1	0	25.0	0
15	100%	4.5	3.3	21.4	0.7
22	0%	5.8	0	24.4	0
Total log inactivation* (sum of A × B or A × C)		3.3		0.7	
Sensitivity† (mJ cm^{-2} per log of inactivation)		4.5		21.4	
Calculated RED (mJ cm^{-2})		15		15	

* The total log inactivation is the weighted average of the log inactivations.
† The sensitivity is based on the UV dose–response curves for each microorganism.

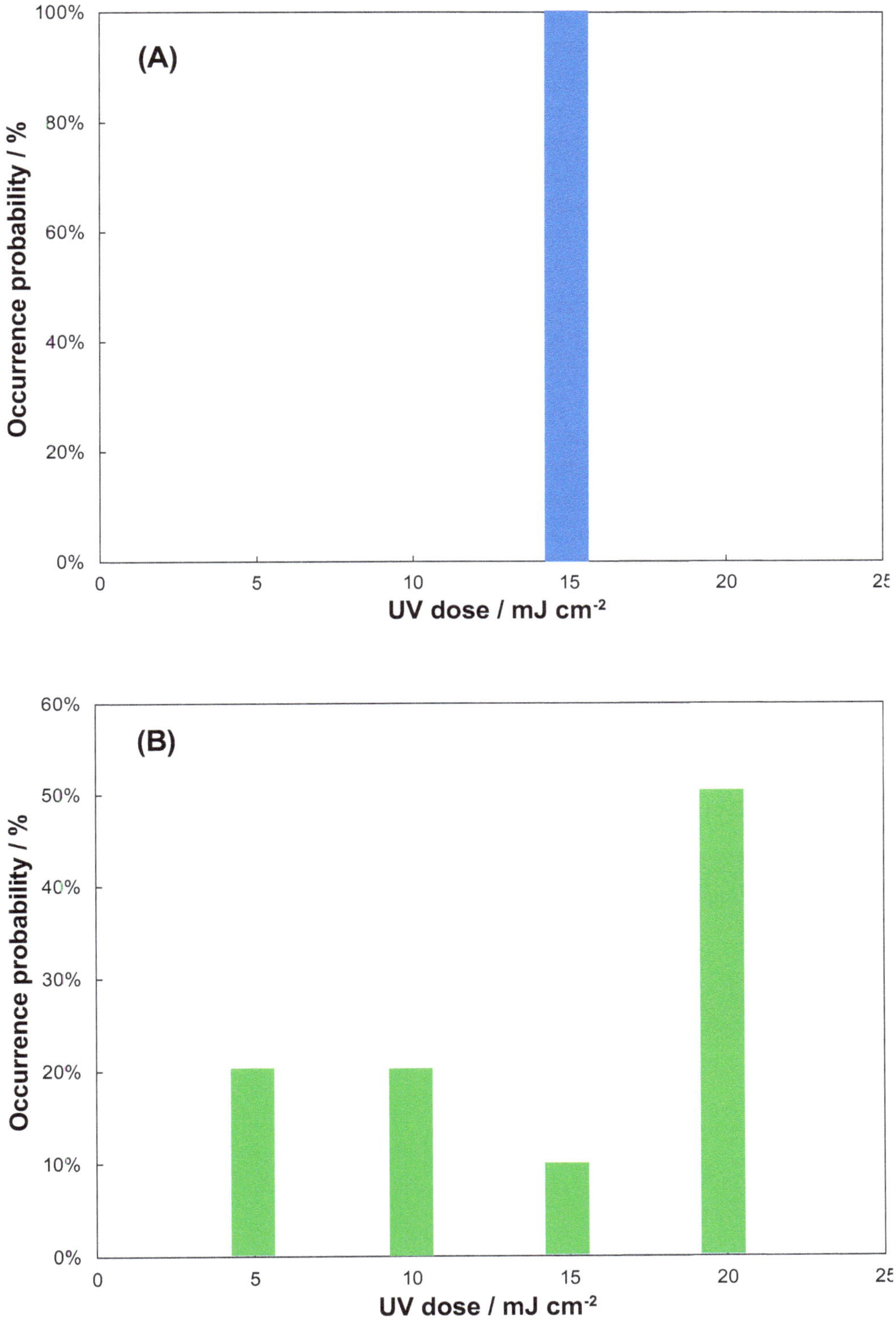

Figure 6-3 Ideal UV reactor UV dose distribution (A); theoretical UV dose distribution (B)

In an ideal UV reactor that only delivers one UV dose (fluence), all of the microorganisms would be exposed to the same UV dose (fluence) (Figure 6-3a), regardless of their path through the reactor. Therefore, the RED would be the same for the target and challenge microorganism (Table 6-1) in an ideal reactor. However, the RED is not typically the same for all microorganisms because the UV dose distribution depends on lamp placement, the UV intensity distribution (which changes with the UVT), and the hydraulics within the reactor. Biodosimetry tests do not provide information on the UV dose distribution within the UV reactor.

To illustrate this, consider a hypothetical UV reactor that delivers UV doses of 5, 10, 15, and 20 mJ cm^{-2} (Figure 6-3b). Based on this UV dose (fluence) distribution, the log inactivation would be 3.0 log for *Cryptosporidium* and 0.7 log for MS2 (Table 6-2).

The UV dose–response characteristics of each of the microorganisms (Figure 6-2) are used to calculate the log inactivation for each UV dose. The log inactivation multiplied by the occurrence probability determines the weighted average total log inactivation. The total log inactivation can then be compared to the UV dose–response characteristics of the microorganism (Figure 6-2) to estimate the RED that would be observed during biodosimetry tests. For this example, the calculated RED would be 12 mJ cm^{-2} for *Cryptosporidium* and 15 mJ cm^{-2} for MS2. In this example, the MS2 RED is approximately 28% larger than the *Cryptosporidium* RED.

Table 6-2 Calculated RED based on non-ideal UV dose distribution and UV dose–response curves

UV Dose (mJ cm^{-2})	Occurrence Probability (A)	Cryptosporidium		MS2	
		Sensitivity† (mJ cm^{-2} Per Log of Inactivation)	Log Inactivation (B)	Sensitivity† (mJ cm^{-2} Per Log of Inactivation)	Log Inactivation (C)
5	20%	2.8	1.8	25.0	0.2
7.5	20%	3.1	2.4	25.0	0.3
15	10%	4.5	3.3	21.4	0.7
22	50%	5.8	3.8	24.4	0.9
Total log inactivation* (sum of A × B or A × C)		3.0		0.7	
Sensitivity† (mJ cm^{-2} per log of inactivation)		3.9		21.4	
Calculated RED (mJ cm^{-2})		11.7		15.0	

* The total log inactivation is the weighted average of the log inactivations.
† The sensitivity is based on the UV dose–response curves for each microorganism.

To calculate the RED bias for the hypothetical UV reactor in Figure 6-3b, the calculated RED of MS2 is divided by the calculated RED for *Cryptosporidium*. For this example, the RED bias would be 1.28. This example RED bias is comparable to the RED biases presented in the UVDGM for 3-log *Cryptosporidium* when the UVT is greater than 95 percent since it is a simplistic UV dose distribution illustration. The UV dose distribution typically becomes broader as the UVT decreases (depending on UV reactor lamp configuration and power), and the RED bias also increases as the UVT decreases.

The UVDGM provides a detailed method to determine the appropriate RED bias during validation; the RED bias calculation here is for illustration purposes only.

The *uncertainty in validation* (U_{VAL}), also referred to as the *experimental uncertainty*, accounts for measurement errors made during validation and uncertainties associated with the equipment installed at the utility. U_{VAL} has between 1 and 3 input variables based on how well the validation testing adheres to the recommended QA/QC limits for the uncertainty of the sensor (U_S), the uncertainty of the fit of the UV dose (fluence)–response curve at a 95-percent confidence level (U_{DR}), the uncertainty of the setpoint (U_{SP}) for UV intensity (irradiance) setpoint control, and the uncertainty of interpolation (U_{IN}) for the calculated UV dose (fluence) control. The UVDGM provides detailed descriptions on how to calculate these uncertainty values. Also, the UVDGM includes two flowcharts for how to calculate U_{VAL}, depending on the UV dose (fluence) monitoring strategy used at the facility. The equations for U_{VAL} for when the sensor uncertainty (U_S) is less than 10 percent (which will be true for the majority of validation data sets) are included below. If the sensor uncertainty is greater than 10 percent, one should consult the UVDGM for U_{VAL} equations that also include sensor uncertainty.

- For the UV intensity (irradiance) set point approach:
  - If U_{DR} is less than 30 percent, $U_{VAL} = U_{SP}$
  - If U_{DR} is greater than 30 percent, $U_{VAL} = (U_{SP}^2 + U_{DR}^2)^{1/2}$
- For the calculated UV dose approach:
  - If U_{DR} is less than 30 percent, $U_{VAL} = U_{IN}$
  - If U_{DR} is greater than 30 percent, $U_{VAL} = (U_{IN}^2 + U_{DR}^2)^{1/2}$

In addition to the RED bias and the uncertainty of validation, UV reactors using MP lamps with nongermicidal sensors should also apply a correction factor to validation test data to account for the *polychromatic bias*. Information on how to derive the correction factor for polychromatic bias is located in Appendix D of the UVDGM (USEPA 2006).

The action spectra correction factor (CF_{AS}) is not included in the UVDGM but is recommended to be included for UV reactors that use polychromatic radiation sources (e.g., MP lamps) to account for the difference in action spectra between the challenge and target microorganism. The Water Research Foundation published guidance on how to calculate the CF_{AS} (Linden et al. 2015), which has been widely used since that date. Another method for CF_{AS} calculation was published by USEPA as an alternative (USEPA 2020). Please refer to these references for detailed methods.

After dividing the RED by the VF and determining the validated UV dose, these validated UV doses are compared to the required UV doses (fluences) (shown in Chapter 4) to determine the inactivation credit achieved. Also, the validated UV dose values can be used to generate relations among the UV dose (fluence), the operating flow rate, the UVT, and the sensor setpoints essential for verifying that the minimum required UV dose (fluence) is delivered by the reactor. These relations can be expressed as performance curves, tables, or an equation where the validated UV dose is a function of the operating conditions, such as flow rate, UVT, and/or number of lamp banks† on.

At the end of this process, a validation report is produced that documents the reactor tested, the test protocol, the validated UV dose values benchmarked to the monitoring system output, and the log inactivation credit that can be achieved for a range of operating conditions (flow rate, UVT, and minimum setpoint value[s]).

MODELING UV DOSE (FLUENCE) DISTRIBUTIONS

Because biodosimetry testing on full-scale UV reactors is very expensive, mathematical models have been developed to aid in the design of UV reactors and to predict performance in addition to undertaking a full-scale biodosimetry test.

Most mathematical models require two modules for a given UV reactor, one to calculate the irradiance (fluence rate) distribution and the other to calculate the UV dose (fluence) distribution. The latter module uses computational fluid dynamics (CFD) concepts. The combination of these two modules allows the prediction of the UV dose (fluence) distribution for a given UV reactor. When this distribution is combined with the inactivation kinetics for a given microorganism, an estimate of the RED can be obtained, which can then be compared with REDs from biodosimetry testing. It should be noted that USEPA does not allow modeling to determine inactivation credit but allows its use to assess hydraulic and other equipment changes or UV reactor performance.

Irradiance Models

Most irradiance (fluence rate) models are based on modeling a linear UV lamp as a sequence of equally spaced point sources (usually 1,000 or more) along the length of the lamp. These are called *multiple point source summation* (MPSS) models. The first such model was introduced by Jacob and Dranhof (1970). Later Scheible et al. (1985) and Suidan and Severin (1986) refined this model as a practical tool for analyzing UV reactors used in wastewater treatment. Finally, Bolton (2000) refined the MPSS model to incorporate the effects of reflection and refraction at the air–quartz–water interface of the quartz sleeve.‡

---

† A *lamp bank* is a group of UV lamps physically connected in a UV reactor.

‡ A commercial program called *UVCalc* is available from Bolton Photosciences Inc. that uses the mathematics of the Bolton (2000) paper with a modification that the lamp is modeled as a series of cylindrical segments. It is called a *multiple segment source summation* (MSSS) model (see *www.uvcalc.com*).

The most sophisticated of these models allow the calculation of the irradiance (fluence rate) for any geometric position in the UV reactor for a specific UVT. The irradiance (fluence rate) distribution in a UV reactor containing MP lamps (polychromatic radiation source) can be modeled by dividing the 200–300 nm spectral range into 20-five-nm bands and performing a calculation for the UVT averaged over each band. These distributions are combined using a weighting factor incorporating the relative photon emission from the lamp, the quartz transmittance, and the germicidal factor specific for that band. This results in a *germicidal* irradiance (fluence rate) distribution.

CFD Models

Computational fluid dynamics is a mathematical model that incorporates the laws of fluid dynamics to predict the actual path followed by a small particle (e.g., microorganism) as it passes through a pipe or a UV reactor. To predict the performance of a UV reactor, the CFD program must be coupled with an irradiance model. First, the CFD program analyzes the UV reactor and possibly connecting pipes and determines an appropriate set of finite volume elements that it needs to do the calculation. The coordinates of the volume elements are sent to the irradiance model program, which then determines the irradiances (fluence rates) for all the volume elements. This information is returned as an array to the CFD program. The CFD program then injects a *particle* (at random) into the reactor stream at a given point and with a starting velocity (speed and direction). As the CFD program determines the path of the particle through the reactor, it *looks up* the irradiance (fluence rate) (from the irradiance distribution array) for each volume element that is traversed and multiplied by the transit time in seconds. This gives the UV dose (fluence) contribution from that volume element. The CFD program then adds these UV dose (fluence) contributions for all the volume elements traversed to obtain the integrated UV dose (fluence) for that particle. The CFD program repeats this process for thousands of particles. The end result is a UV dose (fluence) distribution of frequency of occurrence versus UV dose (fluence).

Figure 3-8 shows two examples of UV dose distributions. The solid curve is from a UV reactor with poor hydraulic mixing. Note that a significant fraction of the microorganisms can pass through the reactor with very low UV doses. This means that such a UV reactor will not be capable of delivering a high log inactivation for pathogens in the water. In contrast, the dotted curve is from a UV reactor with good hydraulic mixing. The distribution is quite sharp and shows that all microorganisms receive almost the same UV dose (fluence). Such a UV reactor will be capable of delivering a high log inactivation.

Once the UV dose (fluence) distribution is known, this is convoluted with the UV sensitivity kinetics for the challenge microorganism (Cabaj et al. 1996, Wright and Mackey 2001) to obtain the RED, which can then be compared with the RED from biodosimetry.

In the modeling exercise, the calculated irradiance (fluence rate) can be averaged over all the volume elements in the reactor to obtain the volume averaged irradiance (fluence rate) $E_{avg}(I_{avg})$. The product of $E_{avg}(I_{avg})$ with the hydraulic residence time (HRT) (in seconds) gives an estimate of the maximum ideal UV dose (fluence) $F_{ideal}(D_{ideal})$ because this calculation assumes perfect radial mixing in the reactor. The ratio of the RED to $F_{ideal}(D_{ideal})$ is called the *mixing efficiency*. The goal of the design engineers of UV reactors is to achieve as high a mixing efficiency as possible.

Examples of Modeling REDs

Several UV equipment manufacturers use mathematical modeling to optimize their UV reactor designs and to make sure that their UV reactors will deliver the required UV doses before undertaking expensive validation procedures. As an example of this type of analysis, Calgon Carbon Corporation§ has provided the following case study.

Calgon Carbon Corporation proposed using their Sentinel 24-in. diameter reactor with five 10 kW medium pressure UV lamps at a WTP with a maximum flow rate of 50 MLD (13.2 mgd) and a design UVT of 93 percent in a filter pipe gallery with very limited space that did not permit the recommended minimum 5 pipe diameters upstream of the UV reactor.

The UVDGM (USEPA 2006) provides guidelines for assessing the impact of installation hydraulics in Section 3.6.2:

> "The inlet and outlet piping to the UV reactor in the UV facility should result in a UV dose delivery that is equal to or greater than the UV dose delivered when the UV reactor was validated."

and

> "CFD modeling and CFD-based UV dose modeling of inlet and outlet conditions may be used to assess whether UV dose delivery at the WTP installation is better than UV dose delivery achieved during validation for given conditions of flow rate, UVT, and lamp output."

The UV reactor was analyzed by CFD with the *UVCalc* irradiance software for a situation with two UV lamps on. With a lamp fouling–aging factor of 80 percent, the mathematical analysis predicted an MS2-derived RED of 22.3 mJ cm^{-2} with the piping configuration in the filter pipe gallery at the designated WTP. The same CFD model was used to predict the performance under the hydraulic conditions as validated at the validation testing center in Portland, Ore., with a predicted RED of 22.2 mJ cm^{-2}. Therefore, CFD is used to show that predicted UV dose delivery is equal to or greater than the UV dose delivered when the UV reactor was validated. The guidelines go on to state that the reliability of the models should be properly evaluated before their results are accepted. This can be achieved by comparing the results predicted by CFD at the validation center with the actual REDs achieved to validate the model. The predicted REDs should generally be within 10 to 15 percent of the actual validation REDs. Furthermore, most governing agencies prefer to see the CFD modeling to show conservative results (i.e., lower REDs predicted) compared to the biodosimetry testing.

---

§ The authors thank Keith Bircher of Calgon Carbon Corporation for providing this example.

In this example, the CFD modeling is used to predict relative performance. That is, it is predicting the result of a small change (influent hydraulics) to show there will be no adverse impact on performance. It is not permitted to use CFD to predict absolute performance, that is, to predict the performance of a UV reactor in the absence of validation testing.

NEW VALIDATION APPROACHES

The UVDGM was published in 2006, and the UV treatment industry has worked to improve validation techniques. The USEPA (2020) published *Innovative Approaches for Validation of Ultraviolet Disinfection Reactors for Drinking Water Systems* to provide clarification on some UVDGM approaches, another CF_{AS} calculation approach, and an alternative validation calculation method.

Specifically, this new document (USEPA 2020) provides guidance for the following:

1. Detailed microbial methods and UV dose–response QA/QC bounds for commonly used microbial surrogates in UV reactor validation;
2. New potential approaches for the development of calculated UV dose monitoring algorithms that eliminate the need for RED bias factors;
3. Approaches for the development of UV dose monitoring algorithms that do not require an online UV transmittance monitor for simplified UV system operations;
4. For UV reactors equipped with medium-pressure UV lamps, implementation of low-wavelength UV sensors and approaches for the development of UV dose monitoring algorithms that account for the disinfection associated with wavelengths below 240 nm;
5. Criteria for the development of validation test matrices, monitoring algorithm goodness of fit and QA/QC requirements, and standardized approaches for defining the validated range of UV reactors;
6. Target UV doses for 4.5, 5.0, 5.5, and 6.0 log inactivation of *Cryptosporidium*, *Giardia*, and virus for UV applications requiring higher levels of disinfection than the maximum 4.0 log provided by the UVDGM; and
7. General validation and data analysis procedures that are commonly implemented in UV reactor validation but are not explicitly documented in the UVDGM; and
8. Modifications to the operating recommendations of the UVDGM to improve the accuracy of UV dose-monitoring with the water treatment application.

Table 6-3 Summary of the USEPA recommendations for hydraulic configurations for validation testing and the UV facility (adapted from USEPA 2006)

Option	Validation Hydraulics	UV Facility Installation Hydraulics
1	The UV reactor is validated with a 90-degree bend directly upstream of the UV reactor. The UV reactor is defined to include a specific amount of straight pipe upstream or downstream of the UV reactor as specified by the UV manufacturer.	The UV reactor should be installed with a minimum of 5 pipe diameters of straight piping between the UV reactor and any upstream hydraulic configuration.*
2	The inlet and outlet configuration is the same as the installation for 10 diameters upstream and 5 diameters downstream of the UV reactor.	Inlet and outlet configuration is the same as when the UV reactor was validated for 10 diameters upstream and 5 diameters downstream of the UV reactor.
3	The velocity at the validation facility is measured at evenly spaced points through a given cross-section of the flow upstream and downstream of the UV reactor.	The velocity at the installation is measured at evenly spaced points through a given cross-section of the flow upstream and downstream and is within 20% of the theoretical velocity determined during validation.

* This approach is not acceptable if the upstream fitting is an expansion or if the upstream valve will be used for flow control. A valve that will be exclusively used for open/close service (e.g., isolation) is acceptable.

VALIDATION EFFECTS ON UV FACILITY DESIGN

It is advantageous to design the UV facility around a particular, previously validated UV reactor. The method, conditions, and location of validation will affect how the UV facility is designed, as described in more detail in this section.

Validation Hydraulics

The inlet and outlet hydraulics of the UV reactor can significantly affect the UV dose delivery; therefore, the USEPA has recommended the strategies presented in Table 6-3 to ensure that the hydraulics at the UV facility are equal to or better than the hydraulics used during validation testing of the reactors. When designing a UV facility, the inlet and outlet piping will need to be designed based on how the UV reactor was validated such that it meets the USEPA recommendations. Whether or not the UV reactor was pre-validated, off-site affects the inlet/outlet piping options for the UV facility.

Off-Site Versus On-Site Validation

UV reactor validation can either occur off-site or on-site. In off-site validation, the UV reactors are pre-validated prior to installation, typically at a third-party validation test center. In on-site validation, the UV reactors are validated at the utility after they have been installed. On-site validation provides more inlet/outlet piping flexibility but also incurs additional design considerations and testing at the WTP. If the selected UV equipment is not pre-validated, the water utility can choose either off-site or on-site validation based on their site constraints and preferences. If choosing between off-site and on-site validation, it is recommended

Table 6-4 Comparison of off-site and on-site validation (adapted from USEPA 2006)

	Advantages	Disadvantages
Off-site	■ A broader range of flow and water quality are tested, so a reactor can be validated for more than one application. ■ Installation hydraulics are general, allowing for installation at most WTPs. ■ The process is simpler, and cost is usually lower. ■ Performance of reactors is known before design and construction of facility.	■ Revalidation may be necessary if site-specific hydraulics and water quality do not fall within the validated ranges. ■ Water quality may not match the installation location, potentially resulting in inefficient operations.
On-site	■ Validation occurs using the exact hydraulics of the installation. ■ Water quality during validation is specific to the installation. ■ Having provisions for on-site validation (e.g., feed and sample ports and static mixers) allows flexibility for future testing to optimize performance.	■ Facility is designed and constructed before reactor performance is verified. ■ Water quality is limited to the highest UVT at the facility during the assigned validation period. ■ Testing logistics can be complex, including isolation of the test reactor and potentially needing a permit to discharge test water. ■ Cost is often higher than off-site validation. ■ Disposing test water may require special permits.

that the governing agency be consulted to confirm that the selected testing location is acceptable.

Manufacturers will likely validate UV reactors over a wide range of flow rates and water quality (e.g., UVT) conditions at off-site testing facilities. The inlet and outlet hydraulic conditions during validation will probably be selected, so the UV reactors can be installed in most WTPs. However, the LT2ESWTR requires that the site-specific installation and operating conditions must fall within the range of conditions used when the installed UV reactor was validated off-site. If the validation conditions do not encompass the utility's design criteria or inlet and outlet piping configurations, the utility may request that the UV manufacturer revalidate the unit off-site under specific testing conditions that closely match those of the proposed installation. Alternatively, on-site validation can be performed. For example, a WTP with a design UVT of 80 percent is evaluating different UV reactors. One UV reactor has been validated off-site between 78 percent and 98 percent UVT; however, another UV reactor was only validated offsite between 82 percent and 98 percent UVT. The latter UV reactor would not meet the WTP's design UVT and should not be considered unless additional validation testing is performed at or below the design UVT of 80 percent.

The advantages and disadvantages of the validation locations are summarized in Table 6-4.

If on-site validation is desired, the UV facility design should be adapted to enable testing, as discussed in Chapter 8. In addition, the discharge of the validation testing water needs to be considered because non-pathogenic microorganism (i.e., challenge microorganism) and UVT modifier (e.g., Superhume) will be needed for testing and should not be distributed.

REFERENCES

Bolton, J.R. 2000. Calculation of Ultraviolet Fluence Rate Distributions in an Annular Reactor: Significance of Refraction and Reflection. *Water Res.*, 34(13):3315–3324.

Cabaj, A.; R. Sommer; D. Schoenen. 1996. Biodosimetry: Model Calculations for UV Water Disinfection Devices With Regard to Dose Distributions. *Water Res.*, 30:1003–1009.

DIN 19294:1.2020. Devices for the Disinfection of Water Using Ultraviolet Radiation — Part 1: Devices Equipped with UV Low-Pressure Lamps — Requirements and Testing. Joint Austrian/German Standard. DIN. https://shop.austrian-standards.at/action/en/public/details/667583/OENORM_M_5873-1_2020_01_01.

DVGW. 2006. *UV Devices for Disinfection in Drinking Water Supply: Part I: Requirements on Properties, Function and Operation; Part 2: Testing of Properties, Function and Disinfection Effectiveness; Part 3: Ports and Sensors for the Radiometric Monitoring of UV Disinfection Devices—Requirements, Testing and Calibration*. DVGW (German Association for Gas and Water), Bonn, Germany.

Jacob, S.M. and J.S. Dranhof. 1970. Light Intensity Profiles in a Perfectly Mixed Photoreactor. *American Institute of Chemical Engineering Journal*, 16:359–363.

Linden, K.G.; H. Wright; J. Collins; C. Cotton; S.E. Beck. 2015. Guidance for Implementing Action Spectra Correction with Medium Pressure UV Disinfection. Water Research Foundation, Denver, CO.

ÖNORM. 2003. Plants for the Disinfection of Water Using Ultraviolet Radiation—Requirements and Testing—Part 2: Medium Pressure Mercury Lamp Plants. ÖNORM M 5873-2. Vienna, Austria: ÖNORM.

ÖNORM. 2001. Plants for the Disinfection of Water Using Ultraviolet Radiation—Requirements and Testing—Part 1: Low Pressure Mercury Lamp Plants. ÖNORM M 5873-1. Vienna, Austria: ÖNORM.

Scheible, O.K.; M.C. Casey; A. Forndran. 1985. *Ultraviolet Disinfection of Wastewaters From Secondary Effluent and Combined Sewer Overflows.* Report No. EPA/600/2-86/005. Springfield, Va.: USEPA, National Technical Information Service.

Suidan, M.T. and B.F. Severin. 1986. Light Intensity Models for Annular UV Disinfection Reactors. *American Institute of Chemical Engineering Journal*, 32:1902–1909.

USEPA. 2020. Innovative Approaches for Validation of Ultraviolet Disinfection Reactors for Drinking Water Systems. https://cfpub.epa.gov/si/si_public_record_report.cfm?Lab=NRMRL&direntryid=328890.

USEPA. 2006. *Validation of UV Reactors. In Ultraviolet Disinfection Guidance Manual for the Final Long Term 2 Enhanced Surface Water Treatment Rule* [Online]. https://nepis.epa.gov/Exe/ZyPDF.cgi?Dockey=600006T3.txt.

Wright, H.B. and E. Mackey. 2001. Theoretical Treatment of UV Reactor Dose Monitoring Strategies. In Proc. of the AWWA Water Quality Technology Conference. Denver, Colo: AWWA.

CHAPTER 7

Effects of Water Quality on UV Disinfection

Water quality can have a major effect on the efficiency and performance of a UV disinfection reactor. Several water quality factors are discussed in this chapter; however, by far the most important is the percent transmittance (UVT or %*T*) of the water because it affects the UV facility size and cost.

UV TRANSMITTANCE

UVT is the most important parameter because it measures the amount UV radiation absorbed by the water itself, rendering it unavailable to inactivate microorganisms. Furthermore, when more UV radiation is absorbed (i.e., a low UVT), more lamps are needed to meet a given UV dose than when a higher UVT is occurring. Typically, UVT at 254 nm is used as the primary water quality design parameter; however, the UVT spectrum (i.e., UVT scan) is also important for UV reactors that use polychromatic radiation sources (e.g., MP lamps) because other wavelengths in addition to 254 nm affect UV disinfection performance.

UVT at 254 nm

For UV disinfection systems with LP or LPHO UV lamps, the most important design parameter is the UVT at 254 nm. It is important that this parameter be understood correctly and measured properly. As explained in Chapter 2, UVT (expressed as a percentage) is defined as

$$\text{UVT} = 100 \times \left(\frac{E^t}{E^0}\right) = 100 \times 10^{-A(254)} \qquad \text{(Eq 7-1a)}$$

where E^0 and E^t are the irradiances incident on and transmitted through the cell, respectively, and $A(254)$ is the absorbance (actually, absorption coefficient) at 254 nm for a 1-cm path length. In cases where the UVT is larger than 85 percent (often the case with many drinking waters), a 5-cm or 10-cm cell should be used to measure the absorbance ($A = a\ell$). The absorption coefficient, or $A(254)$, is obtained by dividing the measured absorbance by the path length ℓ. Since the UVT is defined only for a path length of 1.00 cm, the UVT should be determined from the absorption coefficient a, that is

$$\text{UVT} = 100 \times 10^{-a} \qquad \text{(Eq 7-1b)}$$

The UVT decreases as the absorption coefficient increases. A variety of absorbing components can contribute to the absorption coefficient. A major contributor is total organic carbon (TOC), consisting of humic and/or alginic acids plus other possible components, such as phenols. Generally, the absorption coefficient of drinking waters increases as the TOC increases. Other components can also contribute. Some of these components are listed in Table 7-1.

Table 7-1 Summary of the molar absorption coefficients at 254 nm for components that may be present in drinking water

Compound	Molar Absorption Coefficient (M^{-1} cm^{-1})	Absorption Coefficient (cm^{-1}) for a 1 mg/L Solution	pH
Ammonia (NH_3)*	~0	0	11.5
Ammonium ion (NH_4^+)*	~0	0	7.0
Calcium ion (Ca^{2+})*	~0	0	6.5
Ferric ion [$Fe(OH)^{2+}$]*	4,716	0.084	4.0
Ferrous ion (Fe^{2+})*	28	0.00050	5.0
Hydrogen peroxide (H_2O_2)*	18.7	0.00055	7.0
Hydroxide ion (OH^-)*	~0	0	13.3
Hypochlorite ion (ClO^-)*	65.7	0.0019	10.0
Hypochlorous acid (HOCl)†	59.0	0.0017	5.0
Magnesium ion (Mg^{2+})*	~0	0	6.0
Manganous ion (Mn^{2+})*	~0	0	3.6
Ozone (O_3) (aqueous)*	3,250	0.068	~7
Permanganate ion (MnO_4^-)*	657	0.012	5.5
Phosphate ion species ($H_2PO_4^-$, HPO_4^{2-})*	~0	0	5–9
Sulfate ion (SO_4^{2-})*	~0	0	7.0

Table 7-1 Summary of the molar absorption coefficients at 254 nm for components that may be present in drinking water (*continued*)

Compound	Molar Absorption Coefficient (M^{-1} cm^{-1})	Absorption Coefficient (cm^{-1}) for a 1 mg/L Solution	pH
Sulfite ion (SO_3^{2-})*	16.5	0.00052	9.0
Thiosulfate ion ($S_2O_3^{2-}$)*	201	0.0063	5.0
Zinc ion (Zn^{2+})*	1.7	0.000026	5.0

* Bolton et al. (2001).

† Feng et al. (2007).

When collecting and analyzing UVT samples, the water sample should not be filtered or pH adjusted, which is suggested in *Standard Methods* (APHA-AWWA-WEF 1998) because this could bias the UVT results.

Figure 7-1 shows how the average irradiance and the UV dose are affected by changes in the UVT, with the reference set at a UVT = 90 percent (typical of good drinking water). For example, if the UVT is 70 percent, rather than 90 percent, the UV lamps would have to increase power by a factor of almost 1.6 to achieve the same UV dose.

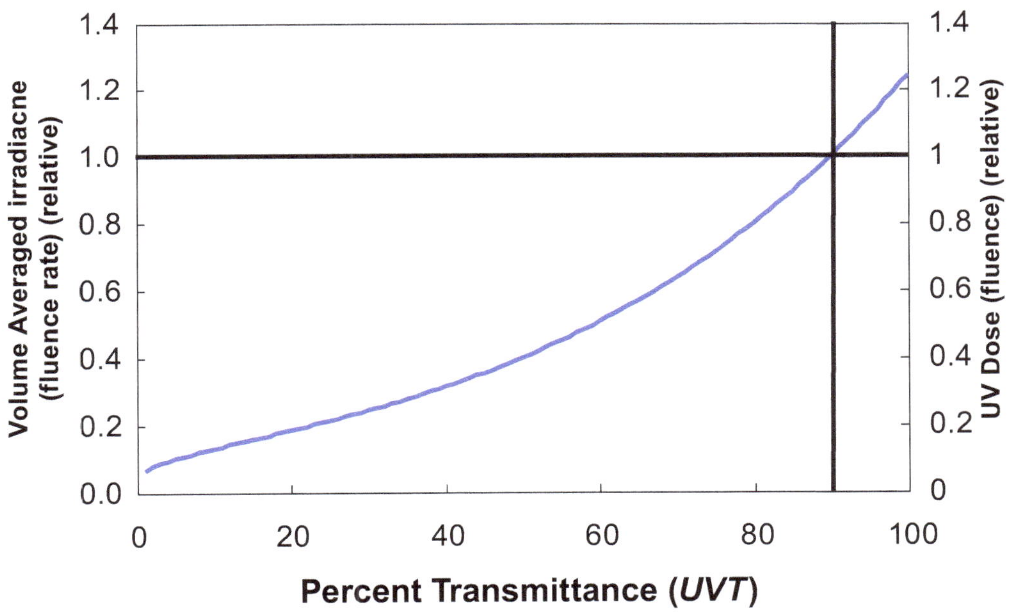

Figure 7-1 Dependence of the relative average irradiance and the UV dose versus the UVT in an ideal UV reactor. The reference is set at a UVT = 90%. The calculations have been made on an annular UV reactor with one UV lamp in the center. The software used was UVCalc, Bolton Photosciences Inc., Edmonton, Alta., Canada (www.uvcalc.com).

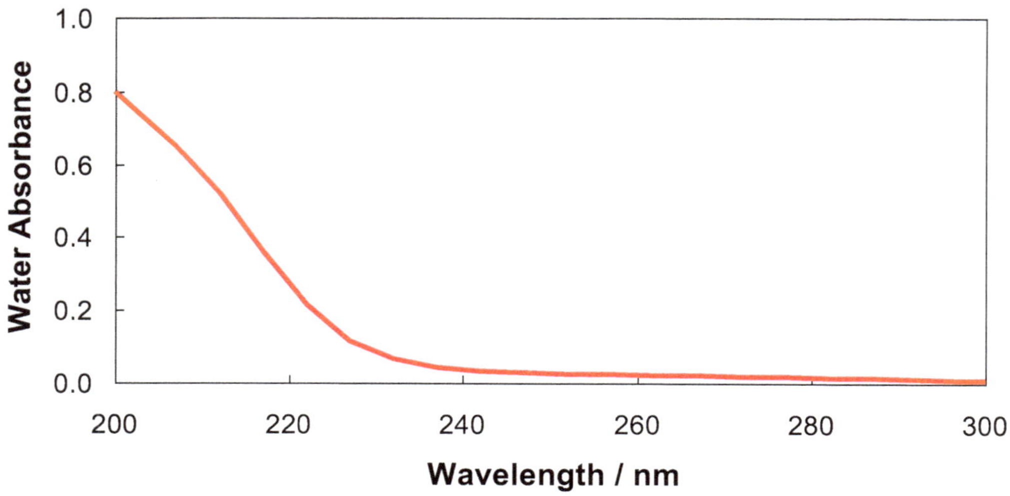

Figure 7-2 Absorbance spectrum of a typical drinking water

UVT Spectrum (200–300 nm)

The dependence of the UV dose on the UVT is different for UV disinfection systems using LP or LPHO UV lamps, as opposed to systems using MP UV lamps. The UV emission from LP and LPHO lamps is essentially monochromatic at 254 nm, whereas the UV emission from MP UV lamps has a broad distribution over the *germicidal* wavelength range of 200–300 nm.

For UV disinfection systems using MP UV lamps, the situation is more complex. Information should be obtained about the emission spectrum of the UV lamps, the UVT spectrum of the water, and the germicidal sensitivity of the target microorganisms. Generally, the dependence of the delivered UV dose on the UVT at 254 nm is qualitatively similar to that in Figure 7-1; however, the quantitative dependence will be a function of the three factors previously listed, and individual calculations should be performed for each specific case.

The absolute absorbance spectrum of drinking water generally increases with the TOC level, although the overall shape does not vary much from one water to another. Figure 7-2 shows a typical drinking water absorbance spectrum. The higher absorbance below 230 nm means that UV radiation in this region (e.g., from an MP lamp) is not very effective from a germicidal standpoint.

FOULING

It is essential that the transmittance of the quartz sleeve between the UV lamp and the water be as high as possible so that the maximum amount of the UV radiation from the lamp can enter the water. However, in some cases, the quartz sleeve can become fouled with deposits that reduce the transmittance and hence adversely affect the performance of the UV reactor.

The UVDGM (USEPA 2006) discusses that the parameters hardness (as $CaCO_3$), alkalinity, temperature, ion concentration, oxidation-reduction potential

(ORP), and pH all may influence the rate of fouling and subsequently, the necessary frequency of sleeve cleaning. Chemicals added immediately upstream of UV disinfection may affect fouling. In addition, changes to treatment chemicals could change the fouling characteristics at an operating UV disinfection facility.

Fouling can occur for the following reasons:

- Compounds for which the solubility decreases as temperature increases may precipitate (e.g., the carbonate, sulfate, and phosphate salts of Mg^{2+}, Ca^{2+}, Fe^{3+}, and Al^{3+}). These compounds will foul sleeves with MP lamps faster than for LP or LPHO lamps because MP lamps operate at higher temperatures.
- Photochemical reactions that are independent of sleeve temperature may cause sleeve fouling (Derrick and Blatchley 2005).
- Compounds with low solubility may precipitate [e.g., $Fe(OH)_3$, $Al(OH)_3$].
- Particles may deposit on the lamp sleeve surface as a result of gravity settling and turbulence-induced collisions (Lin et al. 1999a).
- Organic fouling can occur when a reactor is left off and full of water for an extended period of time (Toivanen 2000).
- Biological fouling on the quartz sleeves, UV reactor walls, and UVT analyzer internal sleeve has been observed in several UV disinfection facilities. A low dose of chlorine (i.e., less than 0.5 mg/L) can be used continuously or intermittently to reduce biological fouling.
- Inorganic constituents can be oxidized and precipitated (Wait et al. 2005) with example shown in Figure 7-3.

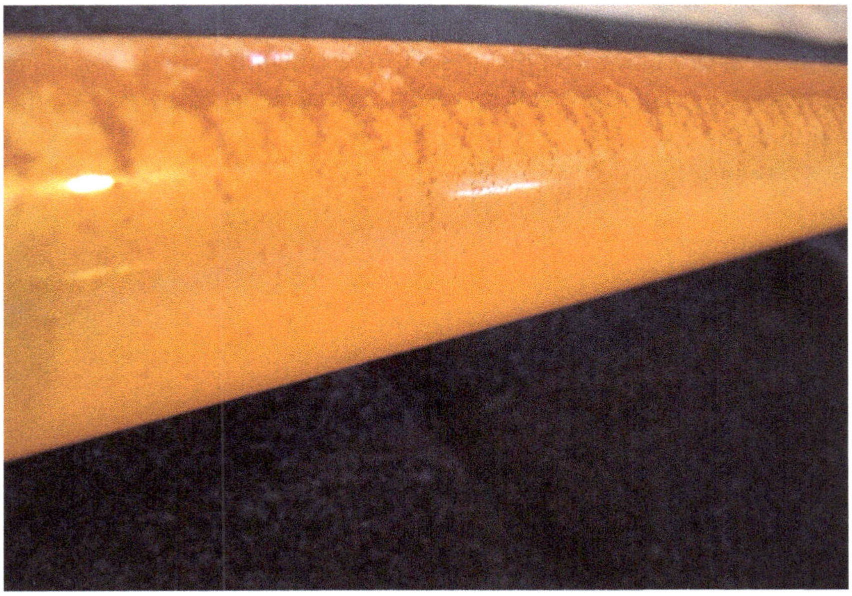

Figure 7-3 Example of fouled sleeve (courtesy of Isaac Wait)

Fouling rate kinetics have been reported as being independent of time following a short induction period (Lin et al. 1999b). Depending on the water quality and UV lamp type, significant fouling may occur in hours or take up to several months.

Gradually, deposits are laid down that occlude the UV radiation and thus decrease the transmittance of the quartz sleeve. To maintain high UV reactor performance, these deposits must be removed periodically.

As noted in Chapter 5, cleaning systems for the removal of quartz sleeve deposits fall into two classes: manual cleaning systems and automatic cleaning systems (see Chapter 5 for more information on cleaning systems).

TURBIDITY

Turbidity is caused by suspended particles in the water. These particles can affect the UV radiation distribution in the water, either by scattering the UV radiation or by absorption of UV radiation by components in the particles. In the case of scattering, the UV radiation is not lost; it is just redirected, and the effect on UV performance is minimal. However, in the case of particle absorption, the UV radiation is lost, and this can adversely affect UV reactor performance. Performance can also be affected by the presence of relatively large particles (1–10 μm in diameter); here, microorganisms can hide behind or within particles and thus be shielded from exposure to the UV radiation. This phenomenon is of particular importance in the UV treatment of wastewater, where such particles are common.

Turbidity is measured in nephelometric turbidity units (ntu). Passantino et al. (2004) have shown that turbidity does not adversely affect UV performance as long as the turbidity is less than 5 ntu. Since most drinking water utilities have filtration systems that keep the turbidity below 0.1 ntu, turbidity is not usually a problem in the UV treatment of filtered water.

Malley (2001) showed that the effects of turbidity on the inactivation of MS2 coliphage depend on the source of the particles (see Figure 7-4). In the case of amorphous wastewater particles (largely organic) or particles from alum floc, the effects are more marked than in the case of filter breakthrough particles (largely inorganic). Nevertheless, for turbidities less than 1 ntu, the effects of turbidity on UV reactor performance can be ignored.

Later, Amoah et al. (2005) showed that at turbidity values less than 10 ntu, the effect of turbidity on the inactivation of *Cryptosporidium* oocysts and *Giardia* cysts is insignificant. Thus, when the inactivation of these protozoa is a concern, the effects of turbidity may be even less significant.

UPSTREAM TREATMENT

Upstream treatment processes can significantly impact UV reactor performance, particularly if they affect the UVT of the water. Indeed, it may be desirable to add a treatment process (e.g., ozone treatment) that will improve the UVT. This will

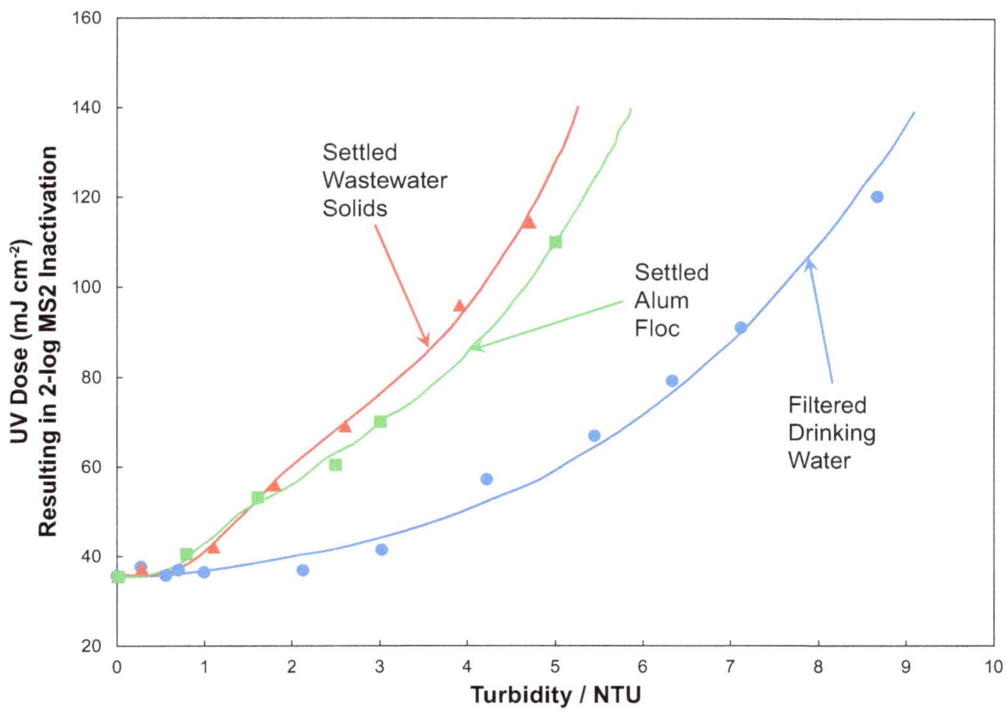

Figure 7-4 Effect of different environmental particles on UV performance (adapted from Malley 2001)

reduce the capital and operating costs of the UV system, and the overall costs may be lower as a result.

Conventional Treatment

Generally, conventional coagulation/sedimentation/filtration processes reduce the TOC of the water, which increases the UVT. As a result, it is more efficient to place the UV system downstream of filtration in the water treatment train, which is required in the LT2ESWTR.

Ozone Treatment

If an ozone treatment system is already in place or one is contemplated, ozone treatment usually increases the UVT. The reason is that ozone adds to double bonds in aromatic rings, which reduces the contributions of these components to the absorbance in the 200–300 nm region. Figure 7-5 illustrates this phenomenon.

There are several water-treatment plants that utilize ozone injection for disinfection credit, and to assist with settleability and filterability of the coagulated water. The ozone treatment significantly increases the UVT even at low ozone doses. For example, one California utility observed an average UVT increase of 4 percent UVT units with an ozone dose of 0.5 mg/L (see Figure 7-6). For utilities with low water UVT, adding ozone treatment and

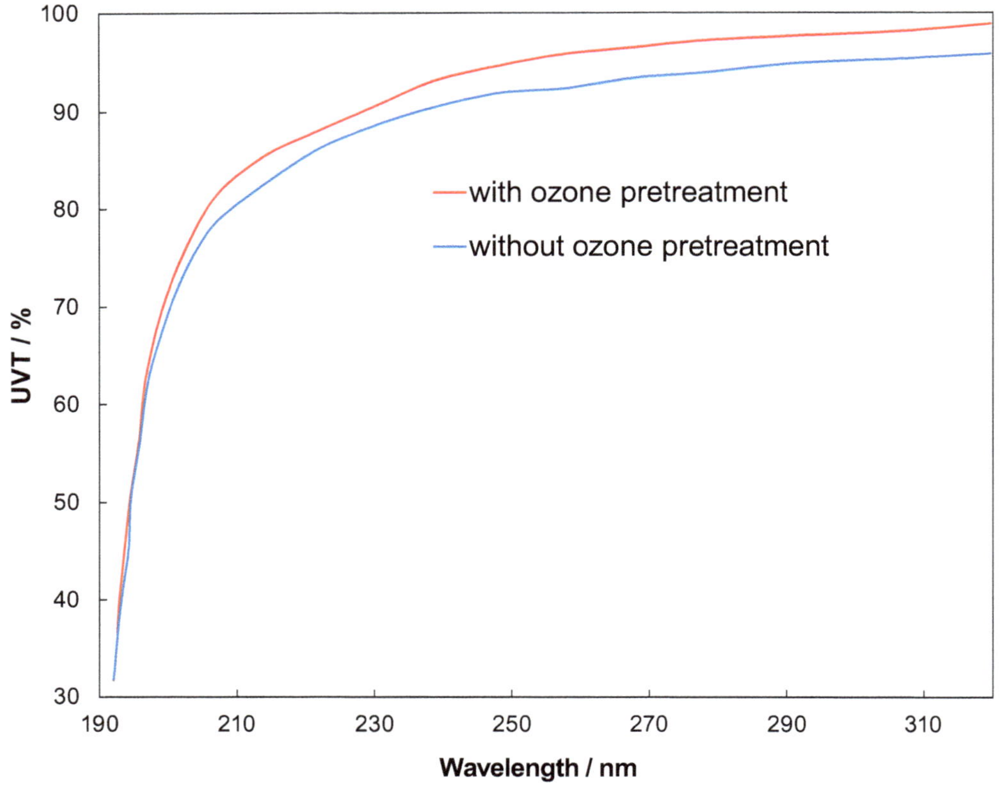

Figure 7-5 Effect of ozonation on the UVT where the ozone is quenched or has decayed prior to the UV disinfection unit (adapted from Malley 2002)

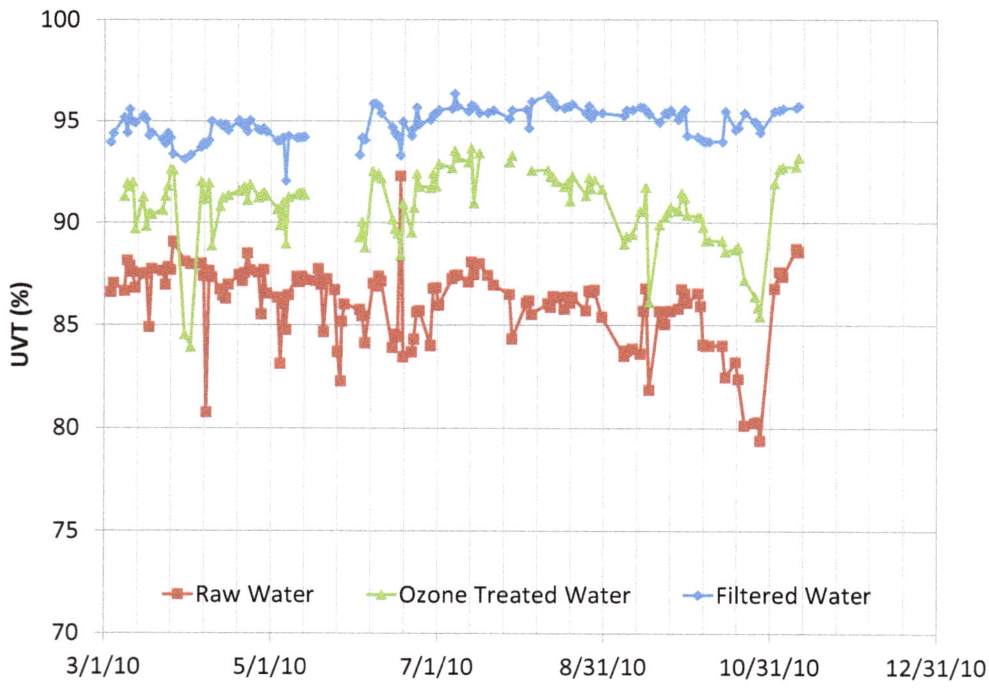

Figure 7-6 Example of UVT increase with an ozone dose of 0.5 mg/L

UV disinfection together may ultimately be less expensive. The UVT increase would be water quality specific and would need to be evaluated to determine the cost-benefit of adding ozone.

CHLORINE

Free chlorine in drinking water exists as a mixture of hypochlorous acid (HOCl) and its conjugate base, the hypochlorite ion (OCl^-). Chlorine can also increase UVT but to a lesser extent than ozone, possibly increasing between 0.5 and 1 percent UVT, depending the chlorine dose and water quality. Because of this, it is recommended that chlorine be avoided as a cleaning agent for UVT analyzers because it can skew the UVT measurement.

In addition, both HOCl and OCl^- undergo UV photolysis (Feng et al. 2007), so if the drinking water has been treated upstream with chlorine (e.g., to oxidize iron and manganese or to keep down growth in filter troughs or settling tanks), the chlorine level will be reduced by a certain amount after the UV system. This phenomenon has been observed in full-scale UV systems (Brodkorb and Richards 2004). Generally, it has been found that a RED of 40 mJ cm^{-2} leads to a reduction in the chlorine level of about 0.1–0.2 mg/L. However, some UV systems operate at higher UV doses where chlorine reduction has been observed to be as high as 1 mg/L.

ALGAE

If algae are present in the water, this may reduce the UVT; however, most algae are removed by the conventional coagulation/sedimentation/filtration processes that exist in most WTPs. Algae growth can be an issue because light from the visible wavelength ranges that can penetrate deep into the upstream or downstream pipe network stimulating photosynthesis in the algae; most UV lamps emit light in the visible range. However, the nutrient levels in most drinking waters are typically too low to make this a significant problem.

BYPRODUCTS FROM UV TREATMENT

Chemical disinfection processes can be expected to produce byproducts, such as THMs and HAAs from chlorine disinfection. However, in the case of UV disinfection treatment, byproduct formation is minimal. Studies have shown that at UV doses up to four to five times those normally used for UV disinfection (about 40 mJ cm^{-2}), there is no significant change in any of the bulk water quality characterization parameters, such as the pH, TOC level, or turbidity.

Chlorine Disinfection Byproducts

Of particular concern are the levels of chlorine DBPs, particularly if the water already contains free chlorine. Fortunately, several studies have found that even at UV doses up to 400 mJ cm^{-2} (10 times higher than normal for UV disinfection), there are no significant changes in the levels of chlorine DBPs (THMs and HAAs) after UV treatment (Malley et al. 1996; Zheng et al. 1999a, 1999b; Shaw et al. 2000; Venkatesan et al. 2003).

Biodegradable Compounds

Since UV radiation is absorbed by components in the water (e.g., organic components of the TOC), some photochemistry is possible, which could produce biodegradable byproducts. However, it is well known that to induce significant photochemical change, such as in UV-driven advanced oxidation processes, UV doses are required at least 10 to 100 times higher than those needed for UV disinfection. Thus, it is highly unlikely that any significant photochemical byproducts could be formed at UV doses needed for protozoan inactivation. This has been confirmed by several groups (Liu et al. 2002, Venkatesan et al. 2003, Malley et al. 1996).

Nitrite

If there is a significant amount of nitrate (NO_3^-) in the water, there can be some photochemical conversion to nitrite (NO_2^-) at wavelengths below 230 nm (Mack and Bolton 1999), where nitrate absorbs. Thus, UV disinfection systems using LP or LPHO UV lamps will not stimulate this conversion. However, MP UV lamps have a significant output below 230 nm. Nevertheless, Sharpless and Linden (2001) found that even under extreme conditions of nitrate at 10 mg/L (the MCL for nitrate in the US), less than 1 mg/L (the MCL for nitrite in the US) of nitrite would be formed. This appears not to be a significant issue, even for MP UV lamps, at least in the US. However, in Europe the MCL for nitrite is 0.1 mg/L as nitrate. As a result, Germany and Austria require a quartz sleeve that blocks UV radiation below 240 nm. Please note that nitrite formation can be a concern for UV advanced oxidation facilities as discussed in the UV Advanced Oxidation Handbook (Collins and Bolton 2016).

SITE-SPECIFIC AND OPERATIONAL ISSUES

Every water source is unique, and there may be some site-specific and operational issues that have to be dealt on a case-by-case basis.

Air Entrainment

In the placement of pipe from the water source or in the use of certain types of pumps, air bubbles may be entrained in the water. This can affect the operation of UV reactors. The air bubbles do not absorb UV radiation, but they reflect it; this changes the irradiance (fluence rate) distribution in the UV reactor, making it more uniform, which is a desirable effect. However, the bubbles will cause the UV sensor readings to drop or become erratic because some of the UV radiation is scattered out of the beam. Air entrainment can also affect the UVT measurement. Thus, the UV system may go into an alarm status, even though an adequate UV dose is being delivered. In addition, air can accumulate at the top of the UV reactor, which can cause alarms and possibly reactor shutdown. For these reasons, it is desirable design the UV facility such that the UV reactors are installed at the hydraulic low point and to manage operations so that air bubbles are not entrained in the water flow. Air release valves at piping high points can be used to help reduce air entrainment.

Reservoir Turnover

In some lakes and uncovered reservoirs, there are seasonal or storm-induced turnovers that introduce a much higher turbidity, lower UVT levels, and higher iron and manganese levels into the water. As previously noted, turbidity levels as high as 1 NTU are not a problem for drinking water UV systems, but significantly higher levels could cause a degradation in UV performance. Higher iron and manganese levels can affect the UVT and fouling of sleeves.

REFERENCES

Amoah, K.; S. Craik; D.W. Smith; M. Belosevic. 2005. Inactivation of *Cryptosporidium* Oocysts and *Giardia* Cysts by Ultraviolet Light in the Presence of Natural Particulate Matter. *Journal of Water Supply: Research and Technology - AQUA*, 54(3):165–178.

APHA, AWWA, and WEF (American Public Health Association, American Water Works Association, and Water Environment Federation). 1998. *Standard Methods for the Examination of Water and Wastewater*. 20th ed. Washington, D.C.: APHA.

Bolton, J.R.; M.I. Stefan; R.S. Cushing; E. Mackey. 2001. The Importance of Water Absorbance/Transmittance on the Efficiency of Ultraviolet Disinfection Reactors. In Proc. of the First International Congress on Ultraviolet Technologies [CD-ROM]. Scottsdale, Ariz.: International Ultraviolet Association.

Brodkorb, T. and D. Richards. 2004. UV Disinfection Design to Avoid Chlorine Destruction in High UVT Waters. Presented at Ontario Water Works Association Conference, Niagara Falls, Ont., Canada.

Collins, J. and J.R. Bolton. 2016. *Advanced Oxidation Handbook*, American Water Works Association, Denver, CO. ISBN 978-1-58321-971-3.

Derrick, B. and E.R. Blatchley, III. 2005. Field Investigations of Inorganic Fouling of UV Systems in Groundwater Applications. In Proc. of the Third International Congress on Ultraviolet Technologies. Scottsdale, Ariz.: International Ultraviolet Association.

Feng, Y.; D.W. Smith; J.R. Bolton. 2007. Photolysis of Chlorine Species (HOCl and OCl^-) With 254 nm Ultraviolet Light. *Journal of Environmental Engineering and Science*, 6:277–284.

Lin, L.-S.; C.T. Johnston; E.R. Blatchley, III. 1999a. Inorganic Fouling at Quartz:Water Interfaces in Ultraviolet Photoreactors I: Chemical Characterization. *Water Research*, 33(15):3321–3329.

Lin, L.-S.; C.T. Johnston; E.R. Blatchley, III. 1999b. Inorganic Fouling at Quartz:Water Interfaces in Ultraviolet Photoreactors II: Temporal and Spatial Distribution. *Water Research*, 33(15):3330–3338.

Liu, W.; S.A. Andrews; J.R. Bolton; K.G. Linden; C. Sharpless; M. Stefan. 2002. Comparison of Disinfection Byproduct (DBP) Formation From Different UV Technologies at Bench Scale. *Water Science and Technology–Water Supply*, 2(5–6):515–521.

Mack, J. and J.R. Bolton. 1999. Photochemistry of Nitrite and Nitrate in Aqueous Solution: A Review. *Journal of Photochemistry and Photobiology C: Photochemistry*, 128:1–13.

Malley, Jr., J.P. 2002. Historical Perspective of UV Use. In Proc. of the AWWA Water Quality Technology Conference. Denver, Colo.: AWWA.

Malley, Jr., J.P. 2001. Engineering of UV Disinfection Systems for Drinking Water. *IUVA News*, 2(3):8–12.

Malley, Jr., J.P.; J.P. Shaw; J.R. Ropp. 1996. *Evaluation of Byproducts Produced by Treatment of Groundwaters With Ultraviolet Irradiation*. Denver, Colo.: AwwaRF.

Passantino, L.; J.P. Malley, Jr.; M. Knudson; R. Ward; J. Kim. 2004. Effect of Low Turbidity and Algae on UV Disinfection Performance. *Journal AWWA*, 96(6):128–137.

Sharpless, C.M. and K.G. Linden. 2001. UV Photolysis of Nitrate: Effects of Natural Organic Matter and Dissolved Inorganic Carbon and Implications for UV Water Disinfection. *Environmental Science and Technology*, 35(14):2949–2955.

Shaw, J.P.; J.P. Malley, Jr.; S.A. Willoughby. 2000. Effects of UV Irradiation on Organic Matter. *Journal AWWA*, 92(4):157–167.

Toivanen, E. 2000. Experiences With UV Disinfection in Helsinki Water. *IUVA News*, 2(6):4–8.

USEPA. 2006. *Ultraviolet Disinfection Guidance Manual for the Final Long Term 2 Enhanced Surface Water Treatment Rule* [Online]. Available: https://nepis.epa.gov/Exe/ZyPDF.cgi?Dockey=600006T3.txt.

Venkatesan, N.; G. Hua; D.A. Recknow; K. Kjartanson. 2003. Impact of UV Disinfection on DBP Formation From Subsequent Chlorination. In Proc. of the AWWA Water Quality Technology Conference. Denver, Colo.: AWWA.

Wait, I.W.; C.T. Johnston; A.P. Schwab; E.R. Blatchley, III. 2005. The Influence of Oxidation Reduction Potential on Inorganic Fouling of Quartz Surfaces in UV Disinfection Systems. In Proc. of the AWWA Water Quality Technology Conference. Denver, Colo.: AWWA.

Zheng, M.; S.A. Andrews; J.R. Bolton. 1999a. Impacts of Medium-Pressure UV on THM and HAA Formation in Pre-UV Chlorinated Drinking Water. In Proc. of the AWWA Water Quality Technology Conference. Denver, Colo.: AWWA.

Zheng, M.; S.A. Andrews; J.R. Bolton. 1999b. Impacts of Medium-Pressure UV and UV/H_2O_2 Treatments on Disinfection Byproduct Formation. In Proc. of the AWWA Annual Conference. Denver, Colo.: AWWA.

CHAPTER 8

Potential Locations for UV Disinfection Facilities

Many water utilities are interested in adding UV disinfection because it is a low-cost method of increasing disinfection levels. This chapter describes potential UV disinfection applications and locations.

FILTERED WATER SYSTEMS

The LT2ESWTR requires some filtered water utilities to apply additional disinfection or removal techniques, depending on the source water's *Cryptosporidium* concentrations. In addition, some filtered systems are considering adding UV disinfection as an additional pathogen barrier. The USEPA regulations concentrate on postfilter applications (USEPA 2006). Therefore, this section describes and compares the postfilter locations where UV disinfection could be applied. If UV disinfection is to be applied *prior* to filtration, the WTP would need to coordinate with their governing agency to determine the requirements for UV disinfection credits.

The hydraulic grade line is one of the most important factors in determining feasible locations because the UV reactors alone have between 6 and 30 in. of head loss, which does not include the additional pipe, valve, and other appurtenance head loss necessary (depending on the location under evaluation).

Combined Filter Effluent

Most WTPs collect the filtered water into a combined header that sends the water to a common disinfection point. The combined filter effluent is one potential location for retrofitting UV disinfection. For this application, the UV reactors are located postfiltration and prior to the clearwell, as shown in Figure 8-1. This type of installation usually requires the construction of a new building to house the UV equipment.

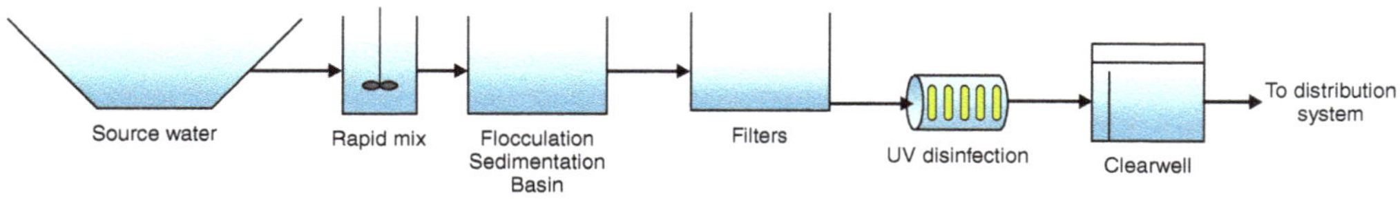

Figure 8-1 Combined filter effluent location in a WTP (USEPA 2006)

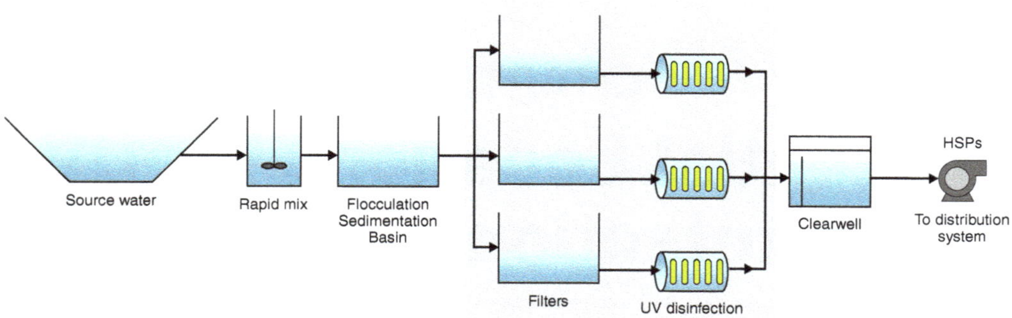

Figure 8-2 Individual filter effluent retrofit example (USEPA 2006)

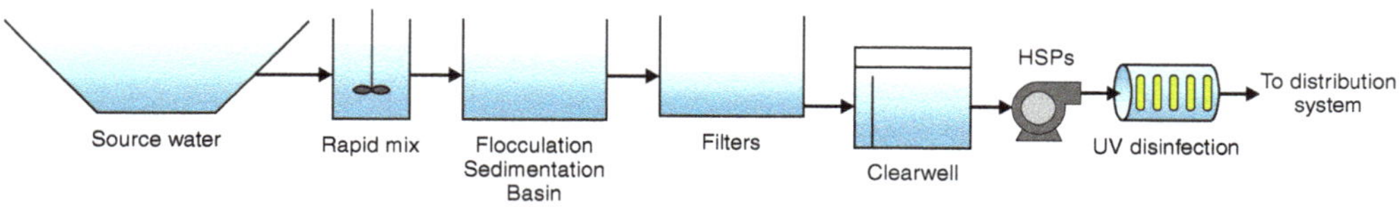

Figure 8-3 Example location of UV facility downstream of the clearwell (USEPA 2006)

Individual Filter Effluent

The second potential installation location is to install the UV reactors on the effluent piping of each filter. In this location, the reactors are typically housed within the existing filter gallery, which eliminates the need for constructing an additional building. Figure 8-2 schematically represents this type of installation.

Combined Filter Effluent Downstream of the Clearwell

The third potential location is after the clearwell and either before or after the high service pumps (HSPs) if used (see Figure 8-3). Either configuration may be advantageous if installation of the UV reactors upstream of the clearwell is not feasible. However, there are significant disadvantages to these options (USEPA 2006) as described in the next section.

Comparison of Filtered Water UV Facility Locations

Many WTPs will only have one or two feasible options for the location of their UV facility. For each option, the utility needs to recognize the advantages and disadvantages before selecting the most appropriate location; the options are summarized in Table 8-1. In order to select the most appropriate location, design criteria, hydraulic constraints, validation approach, footprint estimations, and overall cost need to be evaluated at each potential location.

Table 8-1 Comparison of possible postfiltration UV facility locations

Location	Description	Advantages	Disadvantages
Combined filter retrofit (Figure 8-1)	The UV facility is downstream of the filters and upstream of the clearwell in a separate building. The filtered water is combined prior to UV disinfection.	■ Increased flexibility in the options of system design and operation. ■ Stand-alone operation that is not dependent on the operation of individual filters. ■ Backup disinfection for *Giardia* and viruses may be available upstream or downstream if the UV facility is down. ■ Consistent flow for controls.	■ Increased head loss may adversely affect the operation of the clearwell, reducing storage and/or contact time. ■ May require pumping. ■ May require the use of additional land. ■ May not be feasible with existing site constraints.

Table 8-1 Comparison of possible postfiltration UV facility locations (*continued*)

Location	Description	Advantages	Disadvantages
Individual filter retrofit (Figure 8-2)	A UV reactor is directly downstream of each filter in the filter gallery.	■ Eliminate the need for a new building and additional land. ■ Potentially lower construction costs because a new building is not necessary. ■ Backup disinfection for *Giardia* and viruses may be available upstream or downstream if the UV reactors are down.	■ Increased head loss may adversely affect the operation of filters and/or clearwell. ■ Increased number of UV reactors. ■ Decreased on-line/off-line flexibility if reactors are in-line with filters. ■ Decreased design and operational flexibility and reduced access for maintenance if space is limited. ■ Construction and operational complexity may be increased.

Table 8-1 Comparison of possible postfiltration UV facility locations (*continued*)

Location	Description	Advantages	Disadvantages
Post-clearwell (Figure 8-3)	The UV facility is downstream of the clearwell either upstream or downstream of the high service pumps (HSP).	■ No hydraulic constraint for UV facility if paired with HSPs. ■ May be only retrofit option.	■ Design modifications may be necessary due to suction or high pressure. ■ Backup disinfection is limited. ■ Significant flow variations are typical. ■ Surge suppression equipment may be necessary to eliminate possibility of water hammer. ■ Flow surges and water hammer can increase the risk of sleeve and lamp breaks and mercury releases. ■ May affect HSP operation.

UNFILTERED SYSTEMS APPLICATIONS

Unfiltered systems are also affected by LT2ESWTR, and the WTP may need to provide increased *Cryptosporidium* inactivation, depending on their source water *Cryptosporidium* concentration and current treatment configuration.

Many unfiltered systems have a covered or uncovered storage reservoir included as part of the water system. For covered reservoirs, utilities have the option of installing UV disinfection either on the influent to, or the discharge from, the reservoir (USEPA 2006). For systems with an uncovered reservoir, the UV facility must be downstream of the reservoir to avoid additional treatment requirements per the LT2ESWTR. As with filtered water locations, the most appropriate location depends on hydraulic and space constraints. Water quality issues, such as the following, may also help determine whether the UV installation is most appropriate before or after the storage reservoir:

- Changes in water quality caused by reservoir turnover
- Turbidity spikes from storm events
- Algae counts in the reservoir
- Location of treatment chemicals
- UVT variability prior to and after the reservoir

These water quality issues are described in more detail in Chapter 7.

Along with water quality considerations, the location of the UV facility should also take into account the proximity to HSPs that may create water hammer issues. A surge analysis may be necessary to verify that water hammer will not be an issue. A hydropneumatic tank or pressure-relief valves may be necessary, if water hammer is a concern.

UNCOVERED RESERVOIR APPLICATIONS

As well as requiring additional *Cryptosporidium* treatment, LT2ESWTR also requires systems that have uncovered reservoirs either to cover the reservoir or to treat the discharge from the reservoir. The treatment must provide inactivation and/or removal of 4-log virus, 3-log *Giardia*, and 2-log *Cryptosporidium*.

Some reservoirs are designed such that the inlet and outlet to the reservoir is the same pipe. For these facilities, the UV facility should be designed so that it operates when the water flows from the uncovered reservoir to the customer (USEPA 2006). Because of the high pressures in the reservoir, water hammer may be more prevalent and a surge analysis is recommended. Another hydraulic consideration is how the uncovered reservoir integrates into the distribution system. Most uncovered reservoirs "float" on the system demand, translating to widely variable flows through the UV disinfection facilities. UV facility and UV reactor flows will need to be carefully controlled through the SCADA and PLC programming (see more detail in Chapter 9).

Design elements of this installation are similar to installations downstream of the clearwell, and as such, the items described in the Combined Filter Effluent Downstream of the Clearwell section also apply to this location. The largest uncovered reservoir installation is in Los Angeles, California, and will treat 650 MGD water from the 3.3 billion gallon Los Angeles Reservoir (Figure 8-4).

Lastly, uncovered reservoirs may also experience some of the water quality challenges as described in the Unfiltered Systems Applications section and should be considered.

Figure 8-4 Los Angeles Reservoir UV Disinfection Facility

GROUNDWATER APPLICATIONS

Groundwater systems may consider installing UV disinfection under the following circumstances:

- groundwater under the direct influence (GWUDI) of surface water sources, where the WTP wants to provide an additional disinfection barrier or because of regulatory requirements;
- groundwater system that is required by the GWR to provide virus inactivation. UV disinfection would typically be combined with another treatment technique to meet the 4-log virus inactivation requirement. UV disinfection can be a standalone technique to meet the GWR requirements when UV reactors can be validated for the UV dose required for 4-log virus inactivation; and
- groundwater source where the water utility is interested in providing an additional disinfection barrier.

UV reactors may either be installed at individual wellheads or at a centralized facility, if multiple wells are located in the same area. As with a UV facility at a WTP, an engineering cost analysis can be performed to determine the most appropriate location for a groundwater UV facility. Similar to UV locations near HSPs, groundwater applications should also be evaluated to determine if water hammer is an issue, considering the proximity of the UV reactors to the wellheads.

SMALL SYSTEMS

Small systems (serving less than 3,000 people) may also find UV disinfection an inexpensive method of providing an additional disinfection barrier. The same locations outlined in the previous sections apply. In addition to these locations, small systems may consider installing point-of-entry (POE) UV equipment. POE UV equipment is very small and treats the water before it is distributed, either throughout a small community or at a single building. Small systems UV disinfection reactors may be validated using the NSF/ANSI 55 Ultraviolet Microbiological Water Treatment Systems (2020) as opposed to the LT2EWSTR requirements, and it is up to the governing agency to determine if that is acceptable.

EMERGING APPLICATION FOR NITRIFICATION PREVENTION

Many water suppliers use chloramines for secondary disinfection to comply with the Stage 2 D/DBPR because it reduces DBP formation. However, chloraminated systems are susceptible to nitrification in storage tanks. A fully nitrified tank will have little to no chloramine residual and nitrite formation as ammonia oxidizing bacteria (AOB) reproduce. UV disinfection can readily disinfect bacteria, including AOB. Los Angeles Department of Water and Power (LADWP) conducted a study with Arcadis US, Inc and University of New Hampshire to determine the UV doses necessary to inactivate AOB (Figure 8-5).

For this application, the UV disinfection reactor would be installed on the tank inlet to inactivate AOB prior to entering a storage tank. Some pilot and

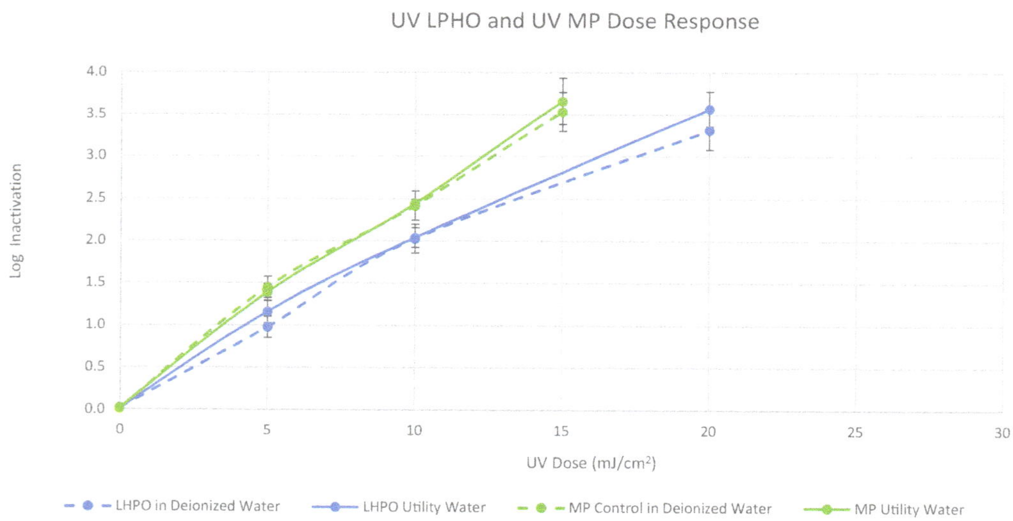

Figure 8-5 Collimated beam testing for AOB inactivation (adapted from So et al. 2019)

full-scale tests of this concept have been completed by East Bay Municipal Utility District (EBMUD) with promising results of less nitrification events (Teefy et al. 2017). Additional studies will be needed to determine the design requirements, costs, and operational considerations of this potential UV disinfection application.

REUSE APPLICATIONS

Dwindling water supplies have caused many water suppliers to consider reuse of the treated wastewater. The treated wastewater, or reclaimed water, can be used for aquifer replenishment or non-potable reuse, groundwater augmentation, and surface water augmentation. UV technologies can be used in any of these reuse applications to accomplish the following:

- microorganism inactivation of bacteria, protozoa, and viruses (except adenovirus) using UV disinfection at UV doses (fluences) required by USEPA;
- direct photolysis of contaminants (e.g., NDMA) using UV radiation at higher UV doses (between 500 and 1,000 mJ cm^{-2}); and
- advanced oxidation of contaminants using UV radiation at high UV doses in conjunction with an oxidant, such as H_2O_2 or chlorine.

These UV applications will be more cost-effective if they are applied to water with the highest UVT and lowest turbidity. The most appropriate locations for UV disinfection or oxidation depend on the specific reuse treatment train. In general, UV disinfection or advanced oxidation should be applied at the point of highest quality water possible and be installed at the back end of the treatment train. However, when installing UV advanced oxidation processes, the residual H_2O_2 concentration should be considered before recycling or blending the water. The AWWA *Advanced Oxidation Handbook* provides more details on these applications (Collins and Bolton 2016).

REFERENCES

Collins, J. and J.R. Bolton 2016. *Advanced Oxidation Handbook*. American Water Works Association, Denver, CO. ISBN: 978-1-58321-971-3.

NSF/ANSI 55. 2020. Ultraviolet Microbiological Water Treatment Systems. NSF International https://webstore.ansi.org/standards/nsf/nsfansi552020.

So, M.; J. Malley, Jr.; C. Cotton; C. Munis; A. Sarkissian. 2019. Non-chemical inhibition of nitrification: A bench and full scale pilot of ultraviolet reactors at the Los Angeles Department of Water and Power, AWWA Water Quality Technology Conference, AWWA, Denver, CO.

Teefy, S.; B. Kawakami; D. Hom; C. Hunt; K. Lorenson. 2019. UV light for control of nitrification in a distribution reservoir, AWWA Water Quality Technology Conference, AWWA, Denver, CO.

USEPA. 2006. *Planning Analyses for UV Facilities.* In: *Ultraviolet Disinfection Guidance Manual for the Final Long Term 2 Enhanced Surface Water Treatment Rule.*

CHAPTER 9
UV Facility Design Issues

This chapter describes the elements of a UV facility design that should be considered in the planning and design process and is summarized from the Planning and Design of UV Facilities chapters in the UVDGM (Chapters 3 and 4) (USEPA 2006).

DISINFECTION GOALS

The specific objectives of a UV facility should be clearly defined before design begins. This will help ensure that the design meets the utility's and the governing agency's expectations based on the regulatory requirements, target microorganism(s), water quality, and the overall disinfection strategy.

Regulatory Requirements

The regulatory requirements for UV disinfection will vary, depending on the water utility's source water *Cryptosporidium* concentrations and current water treatment process. The regulatory requirements are summarized in Chapter 4.

Target Pathogen

The UV facility design will also differ depending on the target pathogen. The UV doses necessary for *Cryptosporidium* and *Giardia* inactivation are lower than those needed to inactivate viruses based on LT2ESWTR UV doses, and as such, the UV facilities may be smaller.

Disinfection Strategy

Utilities typically aim to provide multiple barriers to reduce microbial risk, while minimizing DBP formation (USEPA 2006). A multiple barrier disinfection approach can be accomplished with several different disinfection technologies (e.g., chlorine, chlorine dioxide, ozone, UV disinfection). Of the available options, UV disinfection is typically one of the most cost effective for *Cryptosporidium* inactivation. To a degree, UV disinfection can replace chemicals used to disinfect chlorine-resistant pathogens (e.g., *Cryptosporidium* and *Giardia*), thereby reducing DBP formation. However, the UV doses required for viruses are more than ten times higher because they are based on adenovirus; therefore, UV disinfection is less likely to be used for virus credit compared to the more traditional, chlorine-based disinfection processes. Because of its limited effectiveness at treating viruses and the need to maintain a disinfectant residual in the distribution system, some chlorine-based disinfectant is needed in conjunction with UV disinfection. Water utilities that are changing their disinfection strategies are required by USEPA to prepare a disinfection benchmark (USEPA 2006).

KEY DESIGN CRITERIA

The key design criteria include UVT, fouling–aging factor, flow, and power quality and are described in this section.

UV Transmittance

UVT is the most important water quality parameter for designing a UV facility because the UV facility size depends on the UVT, and the selection of the design UVT is important for both the UV equipment selection and capital and operating costs. Overly conservative design UVT values (i.e., low UVT) can result in the need for additional equipment and a larger building, which increases the overall facility cost. If the equipment is over-designed and does not have the proper turndown capability, overdosing may occur during normal operation, which will result in increased operation costs. However, if the design UVT is inappropriately high, the UV facility could violate the off-specification requirements that are specified in the LT2ESWTR.

The available UVT data should be evaluated to define both the design UVT and the range of UVTs that can be expected. The selected UV reactor should be validated for operation under the range of expected UVTs and flow rates to minimize off-specification operation. It is often useful to evaluate the UVT data in conjunction with flow rate data, if available. The design UVT and the minimum UVT may not be the same value, depending on the flow and UVT data. The design UVT may be higher than the minimum UVT if the low UVT events occur simultaneously with low water demand and flows. Understanding the range of UVTs also helps with equipment selection. If the design UVT and average UVT are significantly different, it may be advantageous to select UV equipment with high turndown capabilities or the ability to turn lamps on and off to optimize power use during higher UVT events.

UVT data is typically reported at a wavelength of 254 nm. If MP UV reactors are being considered, measuring the UVT of the water between 200 to 300 nm (see Figure 7-2) can be helpful. The UV absorbance of water typically increases (i.e., decreasing UVT) at the lower wavelengths in the germicidal range (USEPA 2006).

UV disinfection can be installed in several locations within a WTP (described in Chapter 8). The UVT of the water may change through the treatment process, and UVT measurements should, if possible, be collected at the potential UV locations. If UVT data are not available, sampling is recommended, but the duration of the sampling period and the frequency of sample collection depend on the variability of water quality. For example, a water utility with very stable TOC measurements is likely to have stable UVT measurements and may need only 1 or 2 months of weekly samples. However, a water utility that experiences seasonal changes or utilizes multiple water sources would benefit from more frequent data collection during seasonal events and over a longer period (6 to 12 months or more) under typical source water blends. As previously described, samples should not be filtered or pH-adjusted before measuring the UVT. This technique is contrary to the standard methods; however, filtering the samples could bias the UVT high (USEPA 2006).

Lamp Fouling–Aging Factor

The performance of a UV reactor can be affected by the fouling of the quartz sleeves and UV intensity sensor windows and by lamp and sleeve aging. These factors should be accounted for in the facility design in the fouling–aging factor to minimize the potential for off-specification water.

Sleeve and Sensor Window Fouling

Fouling is typically caused by precipitation of compounds with low solubility, compounds for which the solubility decreases as temperature or ORP increases (e.g., iron compounds), or by photochemical reactions. Fouling rates are dependent on the lamp type and the site-specific water quality and can depend on the ORP, hardness, alkalinity, lamp temperature, pH, and the presence of certain inorganic constituents (e.g., iron, manganese, and calcium) (USEPA 2006) (see also Chapter 7).

These water quality parameters will be helpful for the UV manufacturers to assess qualitatively the fouling potential for their UV reactors and to determine the need for specific cleaning systems. If the data are not available, sampling is recommended. The UVDGM provides guidance on the recommended frequency of sampling. If chemical oxidants must be added upstream of the UV reactor or the water quality analysis indicates the potential for fouling, pilot or demonstration-scale testing can be performed to determine the site-specific fouling rate and cleaning efficiency if deemed necessary.

Lamp, Sleeve, and Sensor Window Aging

UV lamps, sleeves, and sensor windows age as the UV equipment is used, which causes a reduction in UV dose delivery over time. UV lamps have a reduction in

UV radiation output at a rate that depends on the hours of operation, number of on/off cycles, the water temperature, power applied per unit (lamp) length, and heat transfer from lamps (USEPA 2006). Lamp aging curves are typically available from the UV lamp manufacturer. Sleeve and sensor window aging can be caused by prolonged exposure to UV radiation; however, typically this affect is considered to be less of an issue than lamp aging if sleeves and sensor windows are replaced as recommended by the UV manufacturer.

Combined Fouling–Aging Factor Considerations

The fouling–aging factor is site-specific and is based on an assessment of fouling and aging information and typically ranges from 0.4 (NWRI 2003) to 0.9. The fouling factor represents the decrease in the fraction of UV radiation passing through a fouled sleeve as compared to a new sleeve (USEPA 2006). The aging factor represents the decrease in lamp output caused by aging of the lamp or sleeves compared to new lamps and sleeves. Lamp aging evaluations are typically included in the validation report. The fouling–aging factor is calculated by multiplying the fouling factor and the aging factor (USEPA 2006). The UV facility design should specify a fouling–aging factor coupled with a guaranteed lamp life in the design criteria that are provided to the UV manufacturer.

Flow

The key flow-rate design parameter is the maximum flow rate, which will be used to determine the number of UV reactors needed and the maximum flow per UV reactor. Identifying the average and minimum flow rate are also beneficial for selecting the appropriate turndown capacity for the UV reactors to optimize operation costs. The USEPA has recommended several potential methods for determining the design flow rate based on the installation locations described previously (Table 9-1) (USEPA 2006).

Table 9-1 Potential method to determine design flow (USEPA 2006)

Installation Location	Design Flow Basis
Combined filter effluent	Combined rated capacity of all duty filters*
Individual filter effluent	Rated design flow for individual filter
Downstream of the clearwell	Rated capacity of the HSP station
Unfiltered application	Rated capacity of the treatment facility
Groundwater application	Rated capacity of the well pump or well field
Uncovered reservoir application	Rated capacity of reservoir or reservoir outlet

*Does not include redundant filters

Power Quality

UV equipment, depending on the manufacturer, can be highly sensitive to power quality events such as voltage sags/swells or power interruptions. These power quality events can result in the UV lamp losing its arc and having to restart, which would cause disinfection interruption while another UV reactor is energized. The power quality tolerances of UV equipment from three manufacturers are shown in Table 9-2. The impact of power quality events depend on the tolerance of the equipment as well as the time required for the equipment to return to full operation after a shutdown.

Several studies have shown that power quality events experienced by average users are not sufficient to result in a violation of off-specification requirements (Grebe et al. 1996, Cotton et al. 2005). However, USEPA recommends a power quality assessment if the site is (1) known to have power quality problems (e.g., 30 power interruptions and/or brownouts per month), or (2) located in a remote area and the power quality is unknown (USEPA 2006). Existing power supply data may be sufficient to determine if power quality problems exist, but a power quality monitor can be installed if additional information is necessary.

Number of UV Reactors

The seasonal combinations of flow and UVT together will determine the number of UV reactors. However, the permitted WTP flow and maximum validated flow will likely determine the number of UV reactors if the filtered water with high design UVT (>92 percent UVT). A minimum of one redundant UV reactor is recommended, or 10 percent redundancy for larger UV facilities.

Table 9-2 Power quality triggers for UV reactors* (adapted from Cotton et al. 2005) [confirming values with UV manufacturers]

Power Quality Event		LPHO Manufacturer #1	LPHO Manufacturer #2	MP Manufacturer #1	MP Manufacturer #2
Voltage Sag/Swell Tolerance	Voltage†	+/-20%	+/-10%	+/-30%	+/-20%
	Duration‡	2 seconds (s)	>0.03 s	>0.02 s	2 s
Power Interruption Tolerances	Duration§	>0.05 s	>0.03 s	>0.009 s	>0.05 s

* Information shown in the table is compiled from Calgon Carbon Corporation, Trojan Technologies, and Xylem.

† Percent of line voltage. For example, a 10 percent voltage loss is when the voltage is at 90 percent of the line voltage.

‡ 1 cycle is 0.017 s.

§ Power interruption assumes total voltage loss.

HYDRAULICS

The hydraulic design of the UV facility affects the UV reactor performance and hydraulic profile of the treatment facilities. This section describes the hydraulic issues to be considered.

UV Facility Head Loss

Head loss through a UV reactor is manufacturer-specific and can vary from 0.5–3 feet, depending on flow rate and reactor configuration (USEPA 2006). Head loss data should be available for all validated UV reactors and can be provided by the UV manufacturer.

The UV facility will have additional head loss resulting from piping, fitting, and additional equipment. The overall head loss of a UV facility, including the reactor and associated piping, can be between 1 and 8 feet (USEPA 2006). The UV facility headloss is recommended to be calculated to determine how it will affect the WTP hydraulic profile.

Methods for Addressing Head Loss

Changes to the plant design or operation may be required if the head loss through the UV facility is greater than the available head. Potential modifications that could be considered alone or in combination include the following (USEPA 2006):

- eliminating existing hydraulic inefficiencies within the facility to improve head conditions (e.g., replacing undersized or deteriorated piping and valves or sliplining the interior of existing piping with materials having a lower coefficient of friction);
- modifying the operation of the clearwell (e.g., lowering the surface water level);
- modifying the operation of the filters (e.g., increasing the water elevation above the filters or shortening the filter runs);
- installing booster pumps; and
- modifying the operation of the HSPs (only applies if the UV disinfection facility is close to the HSPs).

Inlet/Outlet Piping Design

As discussed in Chapter 6, the UV facility piping design must result in a UV dose delivery that is equal to or greater than the UV dose delivered when the UV reactors are validated. Three UV facility inlet/outlet piping design configurations meet this requirement (based on off-site validation) are summarized in the following list (USEPA 2006). The most appropriate approach depends on site-specific requirements and limitations, and the piping used during validation.

1. **Minimum five pipe diameters of straight pipe upstream of UV reactor.** The UV facility piping has five pipe diameters (plus any deemed necessary during validation) of straight pipe upstream of each UV

reactor. During validation testing, the inlet piping to the reactor consists of either a single 90° bend, an S-bend, or a T-bend, followed by a length of straight pipe if necessary. *This is the most common UV facility design approach.*

2 **Identical inlet and outlet conditions.** Inlet and outlet conditions used during validation match those used at the WTP for at least ten pipe diameters upstream and five pipe diameters downstream of the UV reactor. *This approach is typically used for individual filter effluent UV disinfection installations.*

3 **Velocity profile measurement.** Velocity of the water measured at evenly spaced points through a given cross-section of the flow upstream and downstream of the reactor is within 20 percent of the theoretical velocity with both the validation test stand and the WTP installation (NWRI 2003). The theoretical velocity is defined as the flow rate divided by the cross-sectional area. *This approach has been used for some utility specific hydraulics.*

On-site validation or off-site validation under specific inlet/outlet configurations can be performed if the recommended piping options cannot be met at the UV facility. The options for on-site and off-site validation and how that decision affects the hydraulics of the UV facility are summarized in Figure 9-1. The vast majority of UV disinfection facilities have installed pre-validated UV reactors and have 5 pipe diameters of straight pipe upstream of the UV reactors.

Flow Control and Measurement

The USEPA specifies that flow rate should be controlled and monitored to verify that the UV reactors are within the validated flow limits. Typically, magnetic flow meters are installed in each treatment train.

Flow control can be active or passive. Active flow control uses a flow meter and modulating control valve to regulate the flow through the UV reactors. For passive flow control, equal flow split is not necessary as long as the individual train flows are monitored and within the validated flow range. If equal flow split is desired, it can be accomplished with weirs or submerged orifices. Passive flow control systems lack the ability to control the flow rate through individual reactors and the SCADA controls need to be programmed such that the UV reactors do not exceed the validated flow.

Water Level and Air Release

As previously discussed, drinking water UV reactors are typically closed vessel and designed to be pressurized (i.e., flow full) to reduce the possibility of lamps overheating and for optimal hydraulics. Therefore, the UV facility hydraulic design should have the UV reactors below the downstream hydraulic grade line elevation. In addition, air release valves may be necessary in upstream or downstream piping to release air from the piping. The valve locations will be dictated by the specific configuration of the facility.

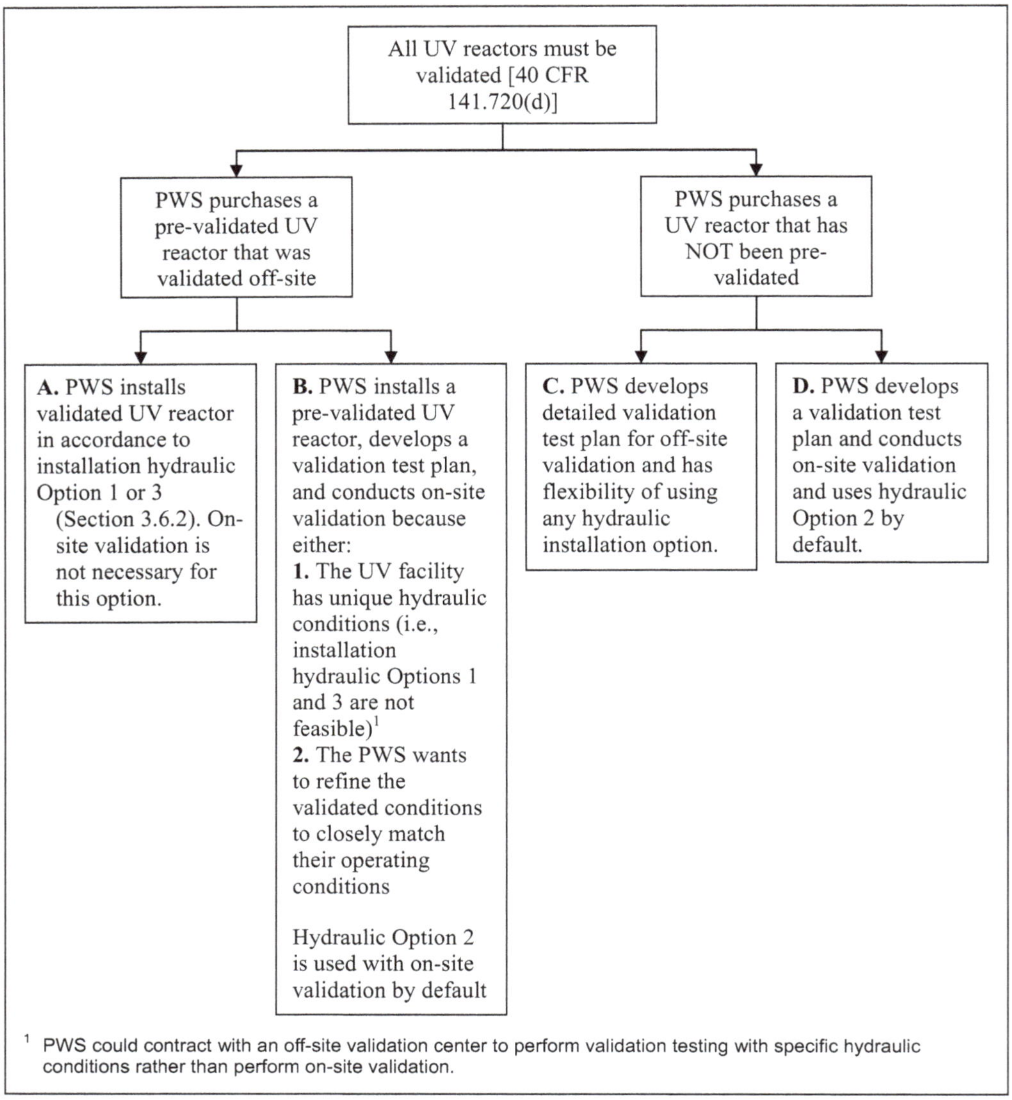

Figure 9-1 UV reactor validation options and how they affect installation hydraulics (USEPA 2006)

ELECTRIC POWER SYSTEMS

Electrical issues to be considered in the design are described in this section. Issues concerning power reliability and backup power are also considered in this chapter.

Power Supply

The electrical power configuration should take into account the power requirements of the selected UV equipment, the disinfection objectives, and power quality issues (USEPA 2006). Excluding high service pumping, the electrical load from UV reactors will typically be among the larger loads at the WTP.

The UV manufacturer should be contacted to determine the power supply voltage and total load requirements for the selected equipment. Multiple power supply voltages may be required, depending on the selected equipment (e.g., 3-phase, 480-volt service for the UV reactors and single-phase, 110-volt service for the UVT analyzer). Some UV equipment can induce current and voltage harmonic distortion that can damage other electrical equipment. The UV facility design and UV equipment should meet the Institute of Electrical and Electronic Engineers (IEEE) 519 Standard to minimize the risk of harmonic distortion.

Backup and Power Conditioning Equipment

As previously discussed, power quality events can result in the UV lamps shutting off, which may result in off-specification operation.

If infrequent but sustained power quality events are expected, a simple backup power supply (e.g., generator) may be sufficient. Existing backup power supplies should be evaluated to determine if they can accept the additional load because of the high load required for UV disinfection.

Power conditioning equipment may be necessary if the power quality events are expected to be frequent and could result in violating the off-specification requirements. Uninterruptible power supply (UPS) systems provide continuous power in the event of a voltage sag or power interruption. The battery capacity supplies power to all connected equipment until a generator starts. There are both on-line and off-line UPS systems to be considered.

SCADA AND CONTROLS

The UV disinfection facility SCADA and controls are critical to maintain compliance with LT2ESWTR, protect the UV equipment, and monitor and report UV disinfection performance.

The UV SCADA and controls will set the number of UV reactors needed online to meet disinfection compliance and will also increase or decrease the number of UV reactors based on the filtered water flow. The controls need to account for the flow variations expected at the UV disinfection facility. For stable flows, the controls could operate with one extra UV reactor to maintain compliance. For variable flows, the controls will need to account for expected flow variability such that the UV reactor flow does not go above the validated flow, which would cause the system to be out of compliance. Another factor is that UV reactors have a maximum flow that will protect from sleeve and lamp breaks, and the UV reactors need to be kept below this threshold.

The SCADA system can also be designed to automatically generate the governing agency's required UV disinfection reports.

SITE CONSTRAINTS/LAYOUT

The following items should be considered when developing the UV reactor and piping configuration and facility layout (USEPA 2006):

- number, capacity, dimensions, and configuration of the UV reactors (including redundancy and connection piping);
- vertical or horizontal orientation of the UV reactor;
- maximum allowable separation distance between the UV reactors and electrical controls if distance limitations apply;
- adequate distance between adjacent reactors to afford access for maintenance tasks (e.g., lamp replacement);
- configuration of the connection piping and the inlet/outlet piping necessary before and after each UV reactor, based on validated hydraulic conditions and UV manufacturer recommendations;
- hydraulic requirements of the overall Water Treatment Plant;
- two-level facilities (if required to meet space or hydraulic constraints) consideration for moving equipment between levels;
- space and piping for booster pumps (if necessary);
- space for electrical equipment, including control panels, transformers, ballasts, generator (if needed), and power conditioning equipment (if needed);
- room and plumbing for a large sink to hand clean the UV sleeves;
- room for storing spare parts and chemicals (if needed);
- lifting capability for heavy equipment; and
- provisions for on-site validation (if applicable), including feed and sample ports, static mixers, space to store tanks for the challenge microorganism and UV absorbing chemical, adequate facilities for laboratory testing, and a place to discharge the treated water.

The UV manufacturer should be contacted to obtain reactor and control panel dimensions to help determine the facility footprint. Also, the footprint and layout will vary, depending on the number and size of reactors and required clearance distances. Therefore, footprints and layouts should be estimated for all of the UV reactors being evaluated. The footprint and head loss through the facility can then be used to evaluate the feasible locations.

When designing an unfiltered UV facility layout, the site layout considerations are generally the same. However, one significant difference is the increased opportunity for debris, which can damage UV reactor components. To address the increased potential for debris, UV facility designs should incorporate features that prevent potentially damaging objects from entering the UV reactor (e.g., screens, baffles, or low-velocity collection areas) if debris is anticipated based on historical experience (USEPA 2006).

When designing a UV facility for groundwater utilities, the most significant differences are access to the site and potential sand particles entering the UV reactor. It may be necessary to enclose the UV facility to limit unauthorized access.

Sand removal equipment (e.g., sand separator) should be installed if the wells are prone to sanding; it may be necessary to install multiple sand removal equipment if significant debris is expected.

COST ESTIMATION

Identifying the most appropriate installation location for UV facilities should consider the projected cost. Preliminary life-cycle cost estimates typically include capital costs and operation and maintenance (O&M) costs. Capital costs include the cost of the UV reactors, building (if necessary), pumping (if necessary), electrical and instrumentation provisions, site work, contractor overhead and profit, validation costs (if necessary), engineering costs, legal costs, and administrative costs (USEPA 2006).

O&M costs account for costs incurred on an annual basis and typically include estimated labor, energy, and equipment replacement costs. The labor, power, and replacement costs will depend on the lamp technology selected, and this cost difference should be accounted for in the O&M costs.

The life-cycle cost analysis may also consider non-monetary preferences that the water utility may have, including (USEPA 2006)

- cost-effectiveness and ability to meet the water system's disinfection and design objectives;
- ease of installation (where applicable);
- operational flexibility and reliability;
- specific maintenance needs; and
- flexibility for future treatment expansion (if applicable).

DESIGN TOOLS AVAILABLE

There are some design tools available to help water utilities navigate through design issues, including the items described in Table 9-3.

MODELING TO ASSIST DESIGN

As noted in Chapter 6, CFD models, in conjunction with irradiance models, are not only able to estimate the RED for comparison with REDs obtained from biodosimetry tests but also produce UV dose distribution curves. These models can be used when the UV facility design has constraints that preclude it from following the UVDGM design recommendations or when new equipment is introduced. Two examples are described below.

- Models can be used to determine if the hydraulic configuration affects UV disinfection performance, especially if the configuration does not meet the UVDGM recommendations (discussed previously in this chapter).

- Models can be used to evaluate whether a minor change in the UV reactor design will affect UV disinfection performance. For example, if a UV manufacturer can no longer purchase the same replacement lamp sleeve as originally validated, a model could show that the new lamp sleeve would provide the same or better disinfection performance. It is likely that this modeling would be done by the UV manufacturer and not the water utility.

Table 9-3 Design tools available

Design Tool	Description	Where to Get It
UVDGM	USEPA's guidance manual on implementing UV disinfection	http://www.epa.gov/safewater/disinfection/lt2/compliance.html
Water Research Foundation reports	More than 32 UV disinfection project reports are available	http://www.waterrf.org/
AWWA Water Treatment Plant Design – UV disinfection chapter	This chapter summarizes design issues for UV disinfection	https://www.awwa.org/Store/Product-Details/productId/28366
CFD modeling	Modeling to evaluate how hydraulic configurations affect UV facility performance	Multiple CFD modeling tools

REFERENCES

Cotton, C.; R. Cushing; D. Owen. 2002. The Impacts of LT2ESWTR Requirements on UV Disinfection Design and Operation. In Proc. of the AWWA Annual Conference. Denver, Colo.: AWWA.

Cotton, C.A.; L. Passantino; D.M. Owen; M. Bishop; M. Valade; W. Becker; R. Joshi; J. Young; M. LeChevallier; R. Hubel. 2005. *Integrating UV Disinfection Into Existing Treatment Plants.* Water Research Foundation.

Grebe, T.E.; D.D. Sabin; M.F. McGranaghan. 1996. *An Assessment of Distribution System Power Quality*. Palo Alto, Calif.: Electric Power Research Institute.

LeChevallier, M. and R. Hubel. 2005. *Integrating UV Disinfection Into Existing Water Treatment Plants*. Denver, Colo.: AwwaRF.

NWRI (National Water Research Institute). 2003. *Ultraviolet Disinfection Guidelines for Drinking Water and Water Reuse*. 2nd ed. Fountain Valley, Calif.: NWRI.

USEPA. 2006. *Ultraviolet Disinfection Guidance Manual for the Final Long Term 2 Enhanced Surface Water Treatment Rule* [Online]. Available: https://nepis.epa.gov/Exe/ZyPDF.cgi?Dockey=600006T3.txt.

CHAPTER 10
Start-Up, Operations, and Maintenance

Unlike traditional chemical disinfectants (e.g., chlorine), UV disinfection, which is a physical disinfection process, does not leave a residual that can be measured to monitor the performance of the disinfection system. This requires significantly different monitoring procedures to determine if the UV equipment is delivering the required levels of inactivation. This chapter describes the start-up, O&M, monitoring, recording, and reporting procedures necessary to confirm that the UV facility is operating properly and in compliance with the regulations required to achieve inactivation credits. The USEPA has recommended several procedures that are not required by the LT2ESWTR, but governing agencies may adopt them as requirements. The governing agency should be contacted to determine any governing agency specific requirements for reactor O&M tasks, reporting needs, and required permitting and submittals. This chapter summarizes the Start-up and Operations (Chapter 6) of the UVDGM (USEPA 2006).

OPERATIONAL REQUIREMENTS

According to the LT2ESWTR, a UV facility must operate within the validated limits of the UV equipment to receive inactivation credit for *Giardia*, *Cryptosporidium*, or viruses. A reactor is operating outside of the validated limits during any of the following conditions (USEPA 2006):

- Flow rate is greater than the highest validated flow rate per UV reactor.
- UVT is lower than the lowest validated UVT.
- UV intensity or validated UV dose is lower than the required intensity or UV dose necessary for the target log inactivation.
- The number of energized lamps is different from the validated number.

- The UV intensity sensor is not calibrated (see monitoring section).
- The UVT analyzer (if used for UV dose calculation) is not calibrated (see monitoring section).

If a UV reactor is operating outside of the validated limits, it is considered to be operating *off-specification*. The LT2ESWTR requires 95 percent of the water delivered to the public each month to be within validated limits; in other words, 5 percent is allowed to be off-specification. The 95 percent value is calculated using the total volume of water treated each month and the volume of off-specification water produced. The off-specification calculation is based on the entire facility and not individual reactors. The validated limits are specific to the UV equipment and the selected UV dose-monitoring strategy. The validation report for the UV reactor should be consulted to determine the reactor specific validated limits.

If the UV intensity setpoint approach is used, the validated limits include flow rate, UV intensity, and the number of energized lamps. If the calculated dose setpoint approach (see Chapter 6) is used, the validated limits include flow rate, UVT, validated UV dose, and the number of energized lamps. With the calculated dose approach, the UV reactor PLC will calculate a UV dose based on the measured flow rate, UVT, UV intensity, and in some cases, the number of lamps energized. The equation used for the UV dose calculation is based on the validation results under these parameters.

A UV reactor can operate at flow rates lower than the lowest validated flow rate, if the UV dose calculation algorithm defaults to the lowest validated flow rate. This is based on the assumption that as the flow rate decreases, the validated UV dose increases. If the flow rate is greater than the validated range, then the reactor is off-specification.

The UV reactor can also operate at UVTs higher than the highest validated UVT if the UV dose calculation defaults to the highest validated UVT. Although forcing the controls to a lower UVT is conservative, this can cause alarms because the UV reactor controls predict lower UV sensor values than are measured. These alarms can be a nuisance if they occur frequently. If the UVT is lower than the validated range, the reactor is off-specification.

Monitoring of select water quality and operational parameters is necessary to determine whether the UV equipment is operating on-specification. In addition to monitoring, proper O&M of UV reactors can reduce the possibility of the UV facility operating off-specification. Routine maintenance of the UV equipment will also decrease the potential for operating outside of the validated limits. Monitoring is discussed in more detail later in this chapter.

STEPS FOR FACILITY START-UP

As the UV facility construction is near completion, several steps should be completed to confirm that the facility meets the LT2ESWTR requirements and any additional governing agency requirements. Start-up activities should include the following (USEPA 2006):

- coordination with the appropriate governing regulatory agency;
- development of a facility O&M manual;
- functional and performance testing;
- spare parts inventory; and
- determination of staffing requirements and training.

Once the UV facility is designed and the governing agency specific requirements are identified, an O&M manual can be developed. The procedures in the O&M manual should minimize the production of off-specification water and maintain operator and public safety. The O&M manual should contain the following items (USEPA 2006):

- general description of UV facility;
- relationship to other unit treatment processes;
- UV reactor design criteria;
- validated operational parameters;
- controls and monitoring;
- compliance monitoring, recording, and reporting;
- standard operating procedures;
- start-up procedures;
- shut-down procedures (manual and automatic);
- safety issues;
- a standard operating procedure for lamp breakage and mercury release;
- emergency procedures and contingency plan;
- alarm response plans;
- preventive maintenance needs and procedures;
- equipment calibration needs and procedures;
- troubleshooting guide;
- equipment component summary;
- spare parts inventory; and
- contact information for equipment manufacturers and technical services.

Prior to the start-up of the UV facility, functional and performance testing should be completed on all of the equipment in the facility. The purpose of the functional testing is to confirm that the specific components operate to the design specifications independently of the other components. Functional testing should include the UV equipment, valves, flowmeters, pumps, instrumentation and control (I&C) systems, flow distribution, and head loss.

Performance testing should be completed to verify that all of the components of the UV facility operate together as intended. The governing agency should be contacted to determine if water produced during performance testing can be sent to the distribution system. Performance testing will not validate the UV reactors but can be used to compare the actual operating condition to those that were validated. The performance testing will also be useful to determine if the requirement of 95 percent of the production water within validated limits can be met reliably. The duration of the testing is site-specific and should demonstrate to the water utility and the governing agency that the facility can meet all applicable requirements during continuous operation. The duration could be as short as one week or as long as six months, depending on the regulatory requirements and the desired scope of evaluation by the water utility.

Performance testing may include the following items (USEPA 2006):

- operation of each UV reactor in automatic mode to verify that the control system is identical to that established during validation testing;
- demonstration of UV reactor start-up and switchover sequences that result from water quality and/or flow rate changes;
- observation of operation, including periods of off-specification operation that arise from alarm conditions and any power quality problems;
- measurement of electrical service voltage, current, and power consumption with different flow and water quality combinations to optimize energy use within the constraints established during validation;
- assessment of the effectiveness of the cleaning system by inspecting sleeve transmittance and condition at regular intervals throughout the test period;
- confirmation that the programmed cleaning frequency correlates with the actual frequency of cleaning;
- confirmation of duty UV sensor accuracy using reference UV sensors;
- observation of ballast temperature and cooling system performance;
- verification of the calibration of the on-line UVT analyzer (if applicable); and
- confirmation of backup generator and/or UPS power transfer to the UV equipment (if applicable).

Before the performance testing is completed, the inventory of spare parts should be checked against the requirements described in the O&M manual. By maintaining the proper inventory of spare parts, off-specification events resulting from the UV equipment can be minimized by having the ability to replace the affected component(s).

The staffing requirements and staff training should also be completed prior to the performance testing. Proper staff training and staffing levels can minimize the production of off-specification water by helping the staff to properly maintain the equipment and be able to react in a timely manner when an off-specification event occurs. Proper staff training is also required for the safety (see Chapter 11) of all staff members. The staff level requirements are site- and UV equipment–specific.

Some utilities may wish to follow performance testing with a short commissioning period to confirm the facility operates as intended, develop and confirm detailed compliance reports, and confirm punch list items are addressed.

ROUTINE OPERATIONS AND MAINTENANCE TASKS

There are no specific regulatory requirements for maintaining UV reactors; however, proper maintenance can reduce the potential for operating outside the validated limits and, therefore, reduce the potential for producing off-specification water. The USEPA-recommended maintenance tasks and their frequencies are found in Table 10-1. The maintenance tasks and the frequency of performing them

Table 10-1 Recommended operations and maintenance tasks* (adapted from Tables 6.3 and 6.4 in USEPA [2006])

Equipment	Frequency	Task
UV reactors	Daily	Perform overall visual inspection of the UV reactors.
		Confirm that system control is on automatic mode (if applicable).
		Check control panel display for status of system components and alarm status and history.
		Review 24-hour monitoring data to confirm that the reactor has been operating within validated limits during that period.
	Monthly	Check reactor housing, sleeves, and wiper seals for leaks and replace any damaged components.
UV lamps	Monthly	Check lamp run time values. Consider changing lamps if operating hours exceed design life and intensity is low.
	Bimonthly (MP lamps) Quarterly (LP and LPHO lamps)	Check intensity of UV lamps. If UV sensors monitor more than one lamp, verify that the lamp with the lowest intensity value is closest to the UV sensor by replacing the lamp closest to the UV sensor with one-fourth of the lamps in each row/bank (minimum of three). Place the lowest intensity lamp next to UV sensor.
	Lamp/manufacturer specific	Replace lamps when any one of the following conditions occurs: - Initiation of low UV intensity or low validated UV dose alarm (UV intensity or validated UV dose equal to or less than setpoint value) after verifying that this condition is caused by low lamp output. - Initiation of lamp failure alarm after verifying it is not a nuisance alarm.
	When lamps are replaced	Send spent lamps to a mercury recycling facility or back to the manufacturer.

Table 10-1 Recommended operations and maintenance tasks* (adapted from Tables 6.3 and 6.4 in USEPA [2006]) (*continued*)

Equipment	Frequency	Task
Ballasts	Daily	Verify that ballast cooling fans are operational and that ballasts are not overheated.
	Semi-annually	Check ballast cooling fans for unusual noise.
	Manufacturer's recommended frequency	Check the ballast cooling fans for dust buildup and damage. Replace if necessary. Replace air filters (if applicable).
UV lamp sleeves	Sleeve/ Manufacturer specific	Replace sleeve when damage, cracks, or irreversible fouling significantly decreases UV intensity of an otherwise acceptable lamp to the minimum validated intensity (or UV dose) level. Inspect sleeves per the cleaning system inspection frequency below for damage and cracks. Adjust the replacement frequency based on operational experience.
Cleaning system	Weekly	Initiate manual operation of wipers (if provided) to verify proper operation.
	Monthly (no cleaning or OCC)	Check cleaning efficiency by recording the UV sensor reading before and after cleaning if fouling is observed.
	Semi-annually (OMC or OMCC)	
	Semi-annually (OMCC)	Check cleaning fluid reservoir (if provided) and replenish as needed. Drain and replace solution if the solution is discolored.
	Manufacturer's recommended frequency	Inspect and maintain OMC or OMCC drive routinely as recommended by the manufacturer.
Duty UV sensors	When duty UV sensors fail calibration	Send the duty UV sensors to a qualified facility (e.g., manufacturer) for calibration, or replace the duty UV sensors.
	Annually	Drain reactor, pull out sensor window, and check for scratches or fouling. Increase frequency if scratches or fouling is observed.
Reference sensor	Annually	Calibrate reference UV sensor by sending it to a qualified facility that uses a traceable standard (e.g., National Institute of Standards and Technology [NIST]).

Table 10-1 Recommended operations and maintenance tasks* (adapted from Tables 6.3 and 6.4 in USEPA [2006]) (*continued*)

Equipment	Frequency	Task
UVT analyzer	Manufacturer's recommended frequency	Clean and replace UVT analyzer parts according to manufacturer's recommended procedure.
Thermometer and/or water level monitor	Manufacturer's recommended frequency	Visually inspect and replace at the manufacturer's recommended frequency.
On-line analyzers and flow meters	Daily	Verify that all on-line analyzers, flow meters, and data recording equipment are operating normally.
Ground fault interrupter (GFI)	Annually	Test trip and maintain GFI breakers in accordance with manufacturer's recommendations.
Valves	Semi-annually	Check operation of automatic and manual valves.

* Maintenance activities should be consistent with manufacturer's instructions.
NOTE: OMC = on-line mechanical cleaning; OMCC = on-line mechanical chemical cleaning; and OCC = on-line chemical cleaning.

can be specific to the UV equipment installed. As such, the UV manufacturer should be consulted when determining the maintenance schedule (USEPA 2006).

MONITORING

To verify that the UV reactors are operating within the validated limits, the LT2ESWTR requires that selected parameters be monitored. Water utilities must at a minimum monitor UV intensity, validated UV dose (if applicable), flow rate, and lamp status. These four items are automatically monitored by the UV reactor or the flowmeter and are not described in detail in this section. In addition to monitoring the selected parameters, it is also required for a water utility to monitor the calibration of the UV sensors, UVT analyzers (if used to calculate UV dose), and off-specification events. Governing agencies may also require additional monitoring (USEPA 2006).

Monitoring UV Sensor Calibration

The USEPA requires that water utilities verify the calibration of duty UV sensors against a reference UV sensor and recommends that this be completed on a monthly basis. Reference sensors are off-line reference sensors that should be at least as accurate as the duty sensors and should be constructed identically to the duty sensors (except for any modifications to make the reference sensor more accurate). The reference sensors should be calibrated on an annual basis by sending the sensor to a qualified facility that uses a traceable standard.

At a minimum, UV sensors in all UV reactors in use (i.e., duty reactors) should be monitored. Monitoring of UV sensors in any stand-by reactors is beneficial because the UV reactor will be operating off-specification if any UV sensors are not in calibration when the reactor is energized. Verifying the calibration of all duty and stand-by UV reactors allows for all reactors to be ready for use, which provides better operational flexibility. Some UV reactor models walk through the UV sensor calibration steps and record the necessary values within its programming.

A duty UV sensor is considered to be in calibration if it reads less than 20 percent higher than the reference sensor (Eq 10-1).

$$\left(\frac{S_{\text{Duty}}}{S_{\text{Ref}}}\right) \leq 1.2 \qquad \text{(Eq 10-1)}$$

where:

S_{Duty} = intensity measured with the duty UV sensor (mW cm^{-2})
S_{Ref} = intensity measured with the reference UV sensor (mW cm^{-2})

Note that Eq 10-1 does not include an absolute value. This allows a UV facility to continue operating if the duty UV sensor reading is more than 20 percent lower than the reference sensor. However, this will result in inefficient operation of the UV reactor because more power will be needed to meet the required UV dose or UV intensity than if a calibrated UV sensor was used.

If the calibration ratio is greater than 1.2, one of the following corrective actions should be taken to avoid the reactor being off-specification:

- Replace the failed UV sensor with a spare calibrated UV sensor.
- Apply a UV sensor correction factor (CF) to the affected UV reactor.

A UV sensor CF is applied to the required UV intensity or required UV dose for the affected reactor(s) and is not applied to the entire UV facility. This corrective action is not energy efficient, but it will allow the facility to continue operation until the failed UV sensor can be replaced. The sensor CF is calculated as the calibration ratio minus 20 percent (Eq 10-2). Twenty percent of the calibration ratio is subtracted to account for the acceptable error in sensor reading previously mentioned.

$$\text{Sensor CF} = \left(\frac{S_{Duty}}{S_{Ref}} - 0.2\right) \qquad \text{(Eq 10-2)}$$

The intensity setpoint or required UV dose should then be multiplied by the sensor CF. If multiple UV sensors in a UV reactor fail the calibration ratio, the highest calculated CF should be applied to the required UV dose or intensity setpoint. The UV sensor CF and the new required setpoint should be reported to the governing agency. The new setpoint will be the basis for determining off-specification operation. Once the UV sensor is replaced, the sensor CF can be removed and the original required UV dose or UV intensity setpoint can be used. Section 6.4.1.1 of the UVDGM provides the detailed procedure for calculating the sensor CF and provides example calculations. The use of sensor CF is not typical but allows for continued operation while a new UV sensor is shipped.

Figure 10-1 provides the procedure for evaluating the duty sensor's calibration and describes the potential corrective actions if calibration errors are found.

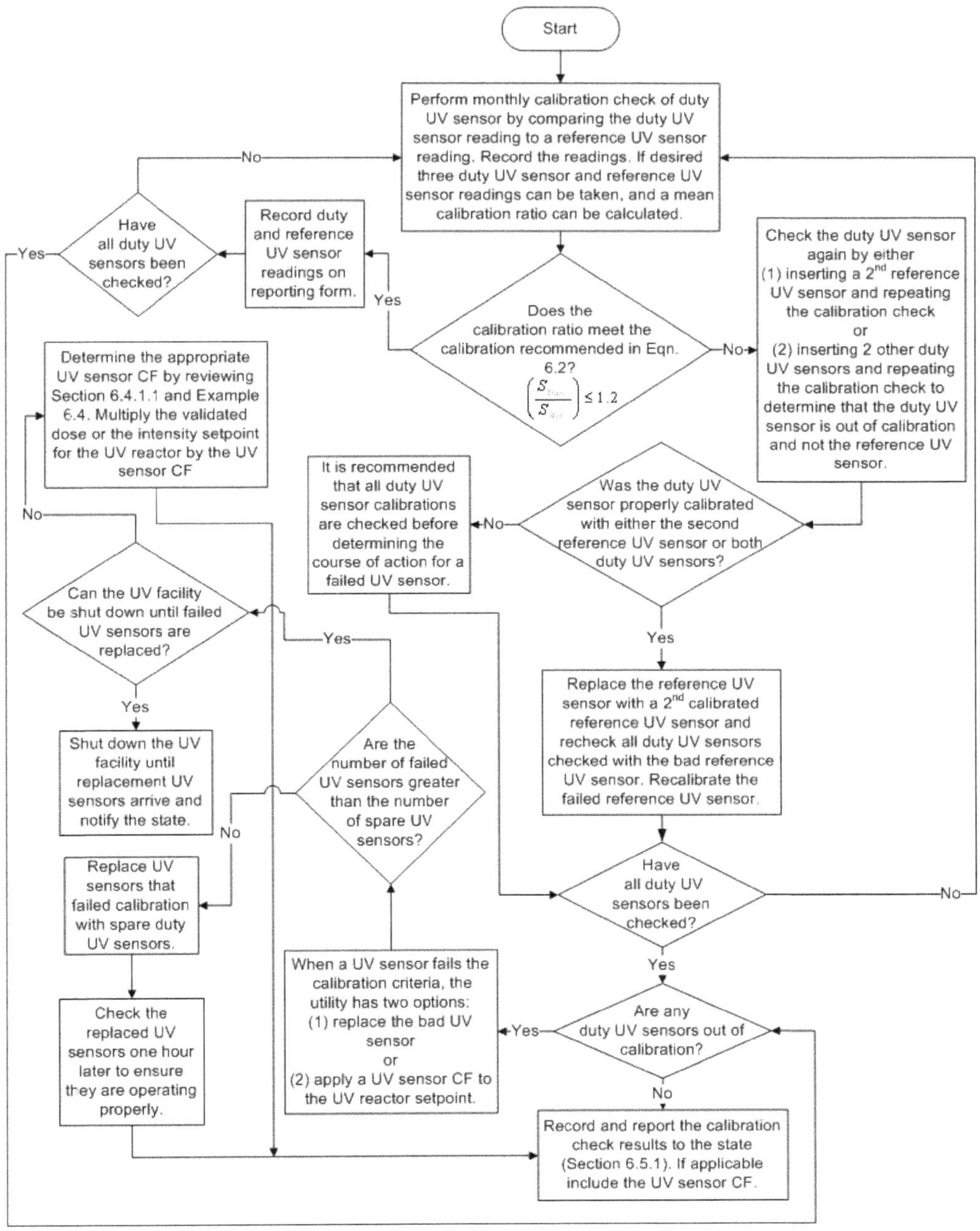

Figure 10-1 Monitoring of UV sensor calibration flow chart (USEPA 2006)

Monitoring the UVT Analyzer Calibration

Monitoring the UVT analyzer calibration is only required when UVT is used to calculate the UV dose during operation (e.g., the calculated dose approach). The calibration of an on-line UVT analyzer is evaluated by comparing the reading of the on-line UVT analyzer to that of a calibrated (according to the manufacturer's specifications) benchtop spectrophotometer using Eq 10-3.

$$\left| UVT_{on-line}(\%) - UVT_{bench}(\%) \right| \leq 2 \text{ percent UVT} \qquad \text{(Eq 10-3)}$$

where:

$UVT_{on\text{-}line}$ = UVT measured by the on-line UVT analyzer (%)
UVT_{bench} = UVT measured by a bench-top spectrophotometer (%)

In this equation, it is important to use the absolute value in the calculations because inaccuracies in the UV dose calculation can result from both conservative and nonconservative errors in UVT measurements. If the UVT analyzer is found to be out of calibration, the analyzer should be recalibrated, or the facility will be operating off-specification. The manufacturer should be contacted for calibration procedures.

The USEPA recommends that on-line UVT analyzers be monitored at least weekly. The monitoring frequency should be adjusted based on the calibration results obtained over the first year of operation and approved by the governing agency. If the UVT analyzer fails calibration in four consecutive weeks, USEPA recommends that the UVT analyzer be monitored on a daily basis. If the UVT analyzer cannot maintain calibration for 24 hours, the water utility should consider one of the following options to minimize the production of off-specification water.

- Enter manual UVT measurements into PLC of all UV reactors every four hours.
- Enter the design UVT value into the PLC and verify daily using a bench-top spectrophotometer that the UVT of the water is less than the design UVT.

These options are not intended for long-term use and should not be used for longer than six months according to USEPA.

UVT analyzers can also be affected by biological fouling on the internal lamp sleeve (as discussed in Chapter 7). Some UVT analyzers can fitted with an auxiliary cleaning chemical to reduce biological fouling. The cleaning solution used should not affect the UVT measurement, and this should be verified prior to full-scale implementation. Please note that sodium hypochlorite should not be used because it can cause UVT values to be falsely high.

Monitoring and Calculating Off-Specification Volume

Water utilities are required to monitor for off-specification events to calculate the volume of off-specification water produced. The following events trigger off-specification operation:

- Reactor is operating outside of validated limits (e.g., flow, UVT, UV intensity, UV dose, number of lamps energized).
- UV sensor is out of calibration, and it is not replaced, or a UV sensor CF is not applied.
- UVT analyzer is out of calibration (if required for UV dose-monitoring strategy) and it is not recalibrated, or remedial actions are not taken.
- UV equipment is not the same or better than the equipment used in validation testing.
- Three methods are available for calculating the off-specification volume produced:
  - A flow totalizer that automatically records off-specification events can be used.
  - The PLC can calculate the volume based on the flow rate in one-minute or less intervals during the off-specification event.
  - The PLC can calculate the volume based on the maximum flow rate during the off-specification event if this method is approved by the governing agency.

These off-specification volumes are divided by the total volume treated to determine the percent off-specification volume produced, which cannot exceed 5 percent. Please note that some state regulations have stricter off-specification limitations.

MONITORING AND RECORDING FREQUENCIES

The LT2ESWTR requires that certain parameters be monitored, but it does not specify the frequency for monitoring. The USEPA has recommended that the validated parameters (e.g., flow rate, UV intensity, validated UV dose [if applicable], and lamp status) be monitored at least every 5 minutes, which is considered to be continuously. These parameters are monitored continuously to verify that the reactors are operating within validated limits, but the measurements only need to be recorded every 4 hours. Recommended monitoring and recording frequencies for required and recommended monitoring parameters are shown in Tables 10-2 and 10-3 (USEPA 2006).

Off-specification alarms should be recorded at a minimum of 5 minute intervals. The off-specification alarm should start as soon as the first off-specification condition is monitored and should continue until the reactor returns to within validated limits. It should be noted that the off-specification event may start before it is first monitored and may end before the reactor is monitored as being within validated limits. USEPA assumed that the underestimation and overestimation of off-specification water will offset over time, thereby minimizing any errors in the calculation of off-specification volume (based on a 5-minute recording frequency). If the monitoring frequency increases, the accuracy of the off-specification calculation also increases (USEPA 2006). Many UV reactor models automatically track and record off-specification events.

REPORTING REQUIREMENTS

Water utilities must prepare and submit monthly reports to the governing agency that detail the percentage of off-specification water produced, the percentage of UV sensors checked for calibration, and the daily low validated UV dose or low UV intensity. The governing agency may also have additional reporting requirements. Example reporting forms that may be adopted by individual states are included in the UVDGM (Section 6.5.2) (USEPA 2006).

The UV facility SCADA and PLC controls can be programmed to automatically develop draft monthly UV disinfection reports. These automatically generated reports should be reviewed and verified prior to submitting to the governing agency.

Table 10-2 Recommended monitoring and recording frequencies for required monitoring parameters (adapted from USEPA [2006])

Parameter	Monitoring Frequency	Recording Frequency
Off-specification alarm	Continuous	Minimum of every 5 minutes until the reactor is on-specification
UV intensity	Continuous	Every 4 hours
UVT*	Continuous	Every 4 hours
Validated UV dose*	Continuous	Every 4 hours
Lamp status	Continuous	Every 4 hours
Flow rate	Continuous	Every 4 hours
Production volume	Continuous	Off-specification events and monthly total
Calibration of UV sensors	Monthly	Monthly
Calibration of on-line UVT analyzer*	Weekly	Weekly†

* Required only if necessary for the UV dose-monitoring strategy (i.e., the calculated dose approach).

† Frequency could be reduced depending on monitoring results.

Table 10-3 Recommended Monitoring and Recording Frequencies for Recommended Monitoring Parameters (adapted from USEPA [2006])

Parameter	Monitoring Frequency	Recording Frequency
Power draw	Continuous	Every 4 hours
Water temperature (Only for MP reactors)	Continuous	Daily
UV lamp on/off cycles	Continuous	Weekly (total cycles per week)
Turbidity (in addition to monitoring otherwise required under subpart H)	Daily	Weekly
pH, iron, calcium, alkalinity, hardness, ORP	Weekly (reduce if fouling is not prevalent)	Weekly
UVT analyzer calibration (if not required for UV dose-monitoring strategy)	Weekly (reduce if appropriate based on operational experience)	Weekly
Operational age* of the following equipment: ■ Lamp ■ Ballast ■ Sleeve ■ UV sensor	Monthly	Monthly
Calibration of flow meter	Monthly	Monthly

* The operational age is the amount of time the equipment was running that month.

TROUBLESHOOTING

An excursion from validated operating limits can be caused by low UV intensity, low validated UV dose, low UVT, high flow rate, poor UV sensor calibration, poor UVT analyzer calibration, or a combination of these conditions. These conditions should be resolved quickly to verify regulatory compliance because they can result in prolonged off-specification operation. Previous sections have provided information for addressing poor UV sensor calibration and poor UVT analyzer calibration. Figures 10-2 and 10-3 provide example procedures that could be followed when the UV intensity/validated UV dose or UVT is approaching the validated limits (USEPA 2006).

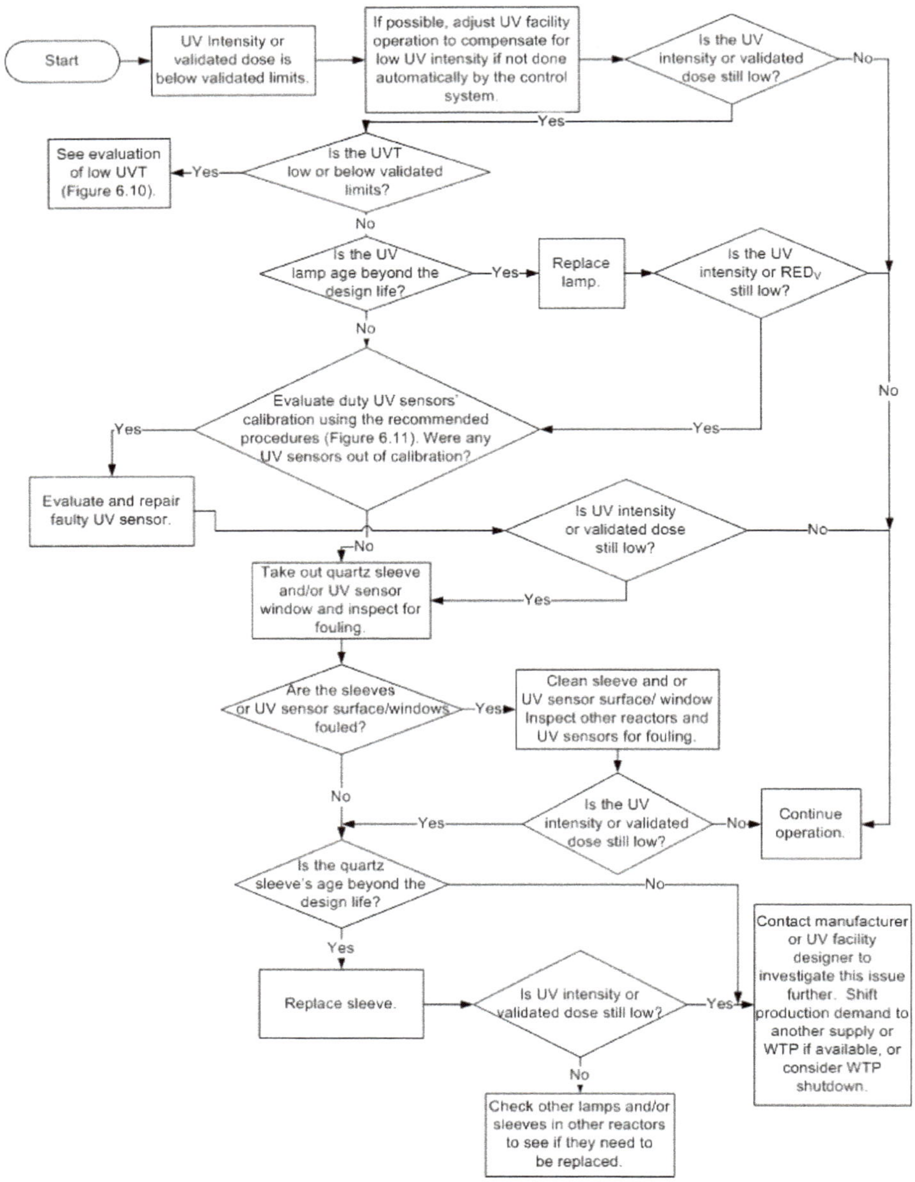

Figure 10-2 Low UV intensity or low validated UV dose decision chart (USEPA 2006)

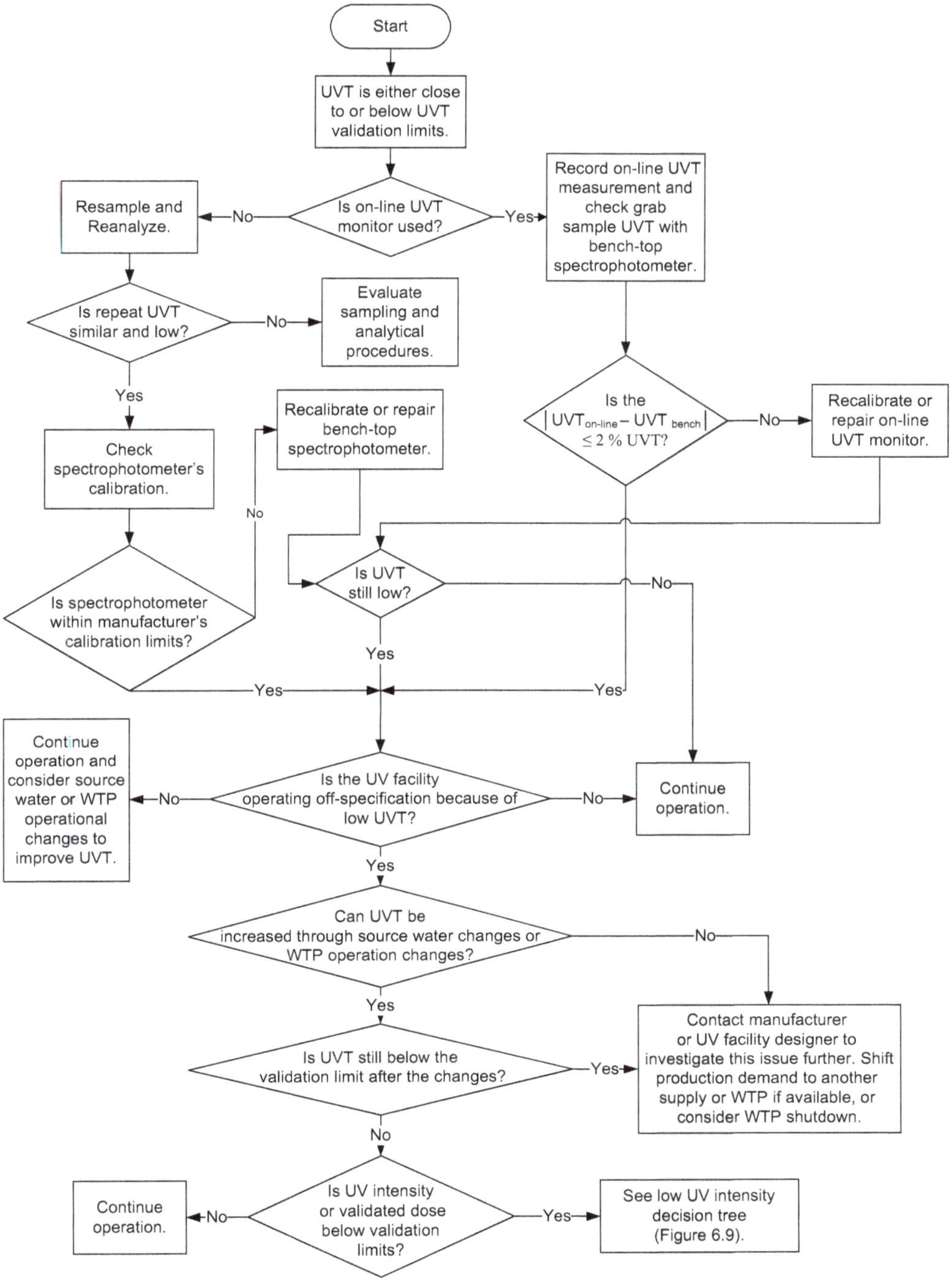

Figure 10-3 Low UVT decision chart (USEPA 2006)

REFERENCE

USEPA. 2006. *Ultraviolet Disinfection Guidance Manual for the Final Long Term 2 Enhanced Surface Water Treatment Rule* [Online]. Available: https://nepis.epa.gov/Exe/ZyPDF.cgi?Dockey=600006T3.txt.

CHAPTER 11
Safety and Handling of UV Equipment

This chapter describes the safety issues associated with handling UV equipment. The Occupational Safety and Health Administration (OSHA) issues regulations and guidance to support operator safety in the workplace, and there may also be specific safety requirements imposed by the governing agency. The following safety issues pertain to UV reactors:

- UV radiation exposure;
- electrical safety;
- burn safety; and
- lamp breakage issues (abrasions, cuts, and mercury exposure).

These items should be included in the UV disinfection facility O&M plan and SOPs. The information in this chapter is summarized from Start-up and Operations of UV Facilities (Chapter 5) and Lamp Break Issues (Appendix E) from the UVDGM (USEPA 2006).

UV RADIATION EXPOSURE

Typical drinking water UV facilities utilize closed-channel reactors that limit the potential exposure to UV radiation. Exposure to UV radiation can occur during reactor maintenance if proper precautions are not employed. There are no enforceable governmental standards for UV radiation exposure. However, for operator safety, the threshold limit value (TLV) for UV radiation exposure should be considered when establishing operations and maintenance procedures. TLVs can change periodically, and *TLVs and BEIs Based on the Documentation of the Threshold Limit Values for Chemical Substances and Physical Agents and*

Biological Exposure Indices (ACGIH 2019) should be referenced to determine the most recent values. The recommended TLVs are dependent on the wavelengths of the emitted UV radiation and the irradiance (mW cm^{-2}) and should be selected based on the installed equipment.

It is not necessary for the UV lamps to be energized during most operations and maintenance tasks. Also, the UV reactors should be designed with safety interlocks that turn off the UV lamps when the reactor is accessed. However, certain maintenance tasks (e.g., UV sensor checks) require the UV lamps to remain energized. Proper personal protective equipment (e.g., UV-resistant face shield) should be worn at all times when the reactor is accessed and while the UV lamps remain energized. If the UV reactors are provided with viewing ports, it should be verified that the viewing window will not permit the transmittance of UV radiation. Warning signs should also be posted in any area where exposure to UV radiation is possible.

ELECTRICAL SAFETY

When accessing the UV reactor or power supply cabinets, UV manufacturer-recommended procedures should be followed. This includes, but is not limited to, disconnecting the main electrical supply and following proper lockout, tag-out procedures. The operator should allow for adequate time (e.g., at least 5 minutes) for the lamps to properly cool down and energy to dissipate before the UV reactor is accessed. All federal, state, and local electrical codes (e.g., National Electric Code [NEC], OSHA) as well as UV manufacturer requirements should be followed, including the following precautions (USEPA 2006):

- proper grounding;
- lockout, tag-out procedures;
- use of proper electrical insulators;
- installation of safety cut-off switches;
- GFI circuits; and
- arc flash analysis and labeling (if needed).

The design of the UV facility should also include proper grounding and insulation of electrical components, which will protect the UV equipment and operators during maintenance tasks.

BURN SAFETY

UV lamps, depending on the lamp technology, may operate at up to 900°C. The UV lamps and sleeves should be allowed to properly cool down before maintenance to minimize the risk of burns. Depending on the lamp, sleeve, and water temperature, the cool-down time for handling may be more than 5 minutes. The electrical equipment (e.g., ballasts) may also become hot during operation and should be evaluated prior to maintenance.

LAMP BREAK ISSUES

UV lamps pose two safety hazards if broken: (1) the lamps and sleeves are constructed of quartz that, when broken, can pose a risk of serious cuts, and (2) UV lamps currently contain a small amount of mercury that can create an inhalation or contact hazard. Operators should be trained in proper clean up procedures in case of a lamp break.

Lamp breaks at a UV facility are divided into two categories: off-line and on-line breaks. Off-line breaks occur when the lamps break when they are not installed in the reactor (e.g., during shipping, handling, cleaning, or storage) or when the reactor is not in operation (e.g., during maintenance). On-line lamp breaks occur when the lamp and lamp sleeve break during reactor operation (USEPA 2006).

Off-Line Lamp Breaks and Prevention Measures

Off-line lamp breaks occur when the UV reactor is not in operation and does not pose a risk to water consumers because the mercury is not in contact with the water (USEPA 2006). As previously described, this category of lamp break may pose a risk to operators in the vicinity of the break as a result of mercury exposure or exposure to quartz fragments and mercury vapor. Off-line lamp breaks are typically caused by handling accidents and can be minimized with proper training and handling procedures. The UV manufacturer should train operators in proper handling and maintenance of UV lamps.

On-Line Lamp Breaks and Prevention Measures

An on-line lamp break is defined as a break that occurs when a lamp and lamp sleeve break while water is flowing through the reactor. These breaks may pose similar hazards to the operators as off-line breaks, as well as risks to water consumers if proper preventative measures are not practiced (USEPA 2006). Table 11-1 summarizes the potential causes of on-line lamp breaks and describes the preventive measures that can reduce each risk.

OPERATOR HEALTH AND SAFETY—MERCURY EXPOSURE LIMITS

OSHA is the regulatory agency that establishes acceptable exposure limits for hazardous chemicals in the workplace. OSHA has established enforceable standards known as permissible exposure limits (PELs) and ceiling PELs (cPELs) for organic mercury compounds. The National Institute for Occupational Safety and Health (NIOSH) also publishes non-enforceable immediately dangerous to life or health (IDLH) concentrations for mercury compounds that can be useful when developing facility safety procedures. Table 11-2 outlines the PELs, cPELs, and IDLHs for mercury compounds and organo-alkyls containing mercury (USEPA 2006).

Proper and prompt lamp break response can typically maintain mercury vapor concentrations below the PELs, cPELs, and IDLHs. If proper clean-up is not followed, the mercury vapor concentration may exceed the limits in areas where mercury can collect (e.g., a drained UV reactor after an on-line break) (USEPA 2006; Table 11-3). The proper clean-up procedures should be outlined in a mercury release response plan (described in the following section).

The mercury vapor concentrations after a lamp break have been observed to be above these levels at several WTPs that had lamp breaks. It is important to have appropriate breathing apparatus when assessing the mercury vapor concentration and potentially while cleaning the area.

MERCURY RELEASE RESPONSE

Response plans for off-line and on-line lamp breaks should be developed. Potential items to be included in these response plans are described in the following section (USEPA 2006). Mercury and materials used during the cleanup procedure are regulated as hazardous wastes and should be disposed of properly.

Off-Line Lamp Break Response

Proper storage and handling of UV lamps will help to minimize the risk of an off-line lamp break. In the event that a lamp break were to occur, a lamp break response plan for containing and cleaning up off-line mercury spills should be available to ensure proper safety procedures are followed.

The size of the mercury release determines the appropriate response to the spill. The USEPA Office of Emergency and Remedial Response recommends that "...[in] the event of a large mercury spill (more than a broken thermometer's worth), immediately evacuate everyone from the area, seal off the area as well as possible, and call your local authorities for assistance" (USEPA 1997). If a lamp break results in a small spill, the mercury can be contained and cleaned up with a commercially available mercury spill kit. Small spills are defined as the amount of mercury in a broken thermometer, or less than 2.25 g (USEPA 1992, 1997). Lamp breaks will typically result in a small spill based on each UV lamp containing between 0.005 and 0.4 g of mercury (USEPA 2006).

Table 11-1 Summary of on-line lamp break causes and prevention methods (USEPA 2006)

Potential Cause	Description	Preventive Measure
Debris	■ Physical impact of debris on lamp sleeves may cause lamp breaks.	■ Installation of screens, baffles, or low velocity collection areas upstream of UV reactors or vertical installation of UV reactors will help prevent debris from entering the reactor.
Lamp orientation	■ Vertical installation relative to the ground may cause overheating and lamp breaks.	■ Install reactors with lamps oriented parallel to the ground to reduce differential heating.
Loss of water flow and temperature increases	■ Lamps may overheat and break. ■ The temperature differential between stagnant water or air and flowing water (on resumption of flow) may cause lamp breaks.	■ Reactors should always be completely flooded and flowing during lamp operation. Temperature and flow sensors that are linked to an alarm and automatic shutoff system can be used to indicate irregular temperature or flow conditions.
Pressure-related events	■ Excessive positive or negative pressures may exceed lamp sleeve tolerances and break the lamp sleeve.	■ A surge analysis should be completed during design to determine the occurrence of water hammer. ■ Pressure relief valves or other measures can be used to reduce pressure surges. ■ Applicable pressure ranges should be specified for lamp sleeves.
Maintenance and handling errors	■ Improper handling or maintenance may compromise the integrity of the lamp sleeve and/or lamp.	■ Operators and maintenance staff should be trained by the manufacturer.

Table 11-1 Summary of on-line lamp break causes and prevention methods (USEPA 2006) (*continued*)

Potential Cause	Description	Preventive Measure
UV reactor manufacturing problems	■ Electrical surges can cause short-circuiting and lamp socket damage.	■ Adequate circuit breakers/ GFI should be specified to prevent damage to the reactor.
	■ Applying power that exceeds design rating of lamps can cause lamps to burst from within.	■ Replacement lamps should be electrically compatible with reactor design.
	■ Misaligned or heat-fused cleaning mechanism may break or damage the lamp sleeve and lamp.	■ Operators and maintenance staff should perform routine inspection and maintenance according to manufacturers' recommendations.
	■ Thermally incompatible materials do not allow for expansion and contraction of lamp components under required temperature range.	■ Designers should specify temperature ranges likely to be encountered during shipping, storage, and operation of lamps to aid the manufacturer in the selection of thermally compatible materials.
	■ UV lamps with manufacturing problems that cause the lamp sleeve to deform at high ballast powers. The deformation causes lamp to explode and potentially sleeve to break.	■ Coordinate with UV manufacturer to determine if the lamp batches purchased has had lamp break issues.

Table 11-2 Health and safety standards for mercury compounds in air (CDC 2021a, 2021b)

Compound	PEL (mg-Hg m^{-3})	cPEL (mg-Hg m^{-3})	IDLH (mg-Hg m^{-3})
Mercury compounds	Not reported	0.1	10
Organo alkyls containing mercury	0.01	0.01	2

Table 11-3 Mercury sampling locations (USEPA 2006)

Media	Location	Purpose
Water	■ Reactor drain ■ Piping downstream of the UV reactor, including the distribution system entry point at a minimum ■ Low velocity areas, such as clearwells	■ Assess the extent of mercury contamination and identify areas requiring cleanup.
Air*	■ Reactor or other locations where mercury vapor may collect ■ Ambient air	■ Wear a breathing apparatus when measuring mercury concentration. ■ Assess whether it is safe to access mercury-contaminated equipment and piping for cleanup. The UV reactor interior may be accessible through an air vent. ■ Assess whether adequate ventilation is provided to safely proceed with mercury cleanup.

* Methods for air sampling are available from the Occupational Safety and Health Administration (OSHA) at http://www.osha.gov/dts/sltc/methods/inorganic/id140/id140.html.

On-Line Lamp Break Response

An on-line lamp break is unlikely; however, an on-line lamp break response plan should be developed prior to operation so that proper response and clean-up measures are followed if an on-line break were to occur. The following is a list of USEPA-suggested components of a lamp break response plan (USEPA 2006). The governing agency should also be consulted when developing the response plans.

- identification of a lamp break;
- safety procedures and equipment;
- site-specific containment measures;
- mercury sampling and compliance monitoring;
- mercury cleaning tools and equipment (e.g., mercury vapor suppressant, mercury spill kit);
- site-specific cleanup procedures; and
- reporting and public notification (if needed).

REFERENCES

ACGIH (American Conference of Governmental Industrial Hygienists). 2019. *TLVs and BEIs Based on the Documentation of the Threshold Limit Values for Chemical Substances and Physical Agents and Biological Exposure Indices*. Cincinnati, Ohio: ACGIH.

CDC (Centers for Disease Control and Prevention). 2021a. CDC—NIOSH Pocket Guide to Chemical Hazards—Mercury compounds [except (organo) alkyls] (as Hg).

CDC. 2021b. CDC—NIOSH Pocket Guide to Chemical Hazards—Mercury (organo) alkyl compounds (as Hg).

USEPA. 2006. Start-up and Operations of UV Facilities, and UV Lamp Break Issues. In *Ultraviolet Disinfection Guidance Manual for the Final Long Term 2 Enhanced Surface Water Treatment Rule*.

USEPA. 1997. *Mercury—Emergency Spill and Release Facts*. EPA 540-K-97-004, OSWER 9378.0-10FS, PB97-963405. Washington, D.C.: USEPA, Office of Emergency and Remedial Response.

USEPA. 1992. *Characterization of Products Containing Mercury in Municipal Solid Waste in the United States, 1970–2000*. Washington, D.C.: USEPA, Office of Solid Waste.

CHAPTER 12
Issues for a Water Utility Manager

This chapter is designed to provide guidance for a water utility manager or water quality manager where the utility is interested in exploring the possible installation of UV disinfection equipment. There are several decisions that need to be made at each stage of the process.

USING AN ENGINEERING CONSULTING FIRM OR DESIGNING IN-HOUSE?

Most large water utilities will procure an engineering consultant to design the UV facility. However, smaller utilities (less than 1 mgd) may prefer an in-house design, depending on their financial situation. The advantage of using an engineering consulting firm is that one can tap into a wealth of expertise concerning UV equipment design and procurement, facility design, safety, and O&M issues. The disadvantage is that it can potentially increase the cost of the project; however, this is not always the case because an engineering consulting firm can potentially save money through careful design and selection of UV equipment.

The advantage of the *in-house design* approach is that one can deal directly with UV equipment vendors and design the UV facility in-house using the utility's engineering staff. There is more control over the process; however, there must be considerable knowledge of the UV disinfection process and the various types of UV equipment and governing regulations. Also, there must be knowledgeable engineers on staff who are capable of managing the total project.

INFORMATION NEEDS FOR UV DISINFECTION EVALUATION

If UV disinfection is being considered, early collection of data will help refine the design and possibly reduce design, construction, and operation costs. The relevant information is

- flow rate (mgd or mL/d) fluctuations to determine the design and average flow;
- complete description of the water treatment plant as it now operates, including water quality data and chemical feeds and doses;
- hydraulic profile throughout existing facilities;
- potential locations in the treatment train of the UV disinfection system—for example, combined filter effluent location;
- UV facility footprint estimations under different UV equipment models;
- monitoring of the percent transmittance (UVT) in a 1-cm path length at 254 nm over at as long a period as possible;
- if MP UV lamps are used, UVT spectrum scans should be measured for the water to be treated from 200 to 300 nm, and a history of these scans should be kept (preferably weekly) for as long a period as possible, especially if virus credit is being sought; and
- other water quality parameters including:
  - *Cryptosporidium* detections in the source water;
  - possible foulants (e.g., calcium, alkalinity, hardness, iron, manganese, pH, ORP);
  - turbidity, particle content, and algae (if UV disinfection will treat a source other than filtered water);
  - TOC levels;
  - any micropollutants of concern (e.g., pesticides, herbicides, pharmaceutical products).

INFORMATION TO PROVIDE AND QUESTIONS FOR A POTENTIAL ENGINEERING CONSULTING FIRM

If an engineering consulting firm will be hired, a facility description package should be compiled, which should include the information gathered (described in the previous section).

The possible engineering consulting firms should be approached either with a request for qualifications (RFQ) or request for proposals (RFP), with the procurement details that meet the local requirements. Local procurement codes will dictate most of the submittal requirements; however, the following are potential, specific UV disinfection elements of the solicitation:

- What experience does the firm have in designing and constructing UV disinfection facilities?
- What is the previous work experience with the utility?
- What is the firm's UV disinfection regulations experience locally and nationally?
- What is the firm's experience with validation of UV reactors?
- Provide at least three references regarding previous UV disinfection projects.
- Provide hydraulic expertise for WTP retrofits.

QUESTIONS FOR A POTENTIAL UV EQUIPMENT MANUFACTURER

If the *in-house design* option has been chosen, the utility will deal directly with UV equipment manufacturers. The utility should develop a UV equipment specification that will be provided to the potential UV manufacturers. The UV specification could include the items described in Table 12-1.

A list of qualified UV vendors should be prepared using criteria such as having validated UV reactor and previous installations of UV disinfection systems for water treatment plants similar in size to the utility. Consult the IUVA UV Buyer's Guide (https://uvsolutionsmag.com/buyersguide/services/0_DrinkingWater__UVSystems) for manufacturers of UV systems for drinking water treatment.

GOVERNMENT REGULATIONS

The relevant government (e.g., USEPA) regulations should be consulted as described in Chapter 4. The state or provincial regulator should be contacted early in the planning process and then strategically throughout the project. Consult http://www.epa.gov/safewater/dwinfo/index.html for a state regulator office in the US. In Canada, consult the regulator office in each province or territory.

Table 12-1 Possible content for UV equipment specifications (adapted from USEPA 2006)

Item	Specification Content
Flow rate	Maximum, minimum, and average flow rates should be clearly identified. The maximum flow rate must be within the validated range documented in the validation report. The minimum flow rate may be important to avoid overheating with MP reactors.
Target pathogen(s) and log inactivation	The log inactivation for the target pathogen(s)

Table 12-1 Possible content for UV equipment specifications (adapted from USEPA 2006) (*continued*)

Item	Specification Content
Required UV dose	The required UV dose for the target microorganism and log inactivation must be verified by the validation process. Additional detail is provided in Chapter 6.
Water quality and environment	The following water quality criteria should be included: – influent temperature – turbidity – UV transmittance at 254 nm – UVT scan from 200–300 nm (MP reactors only) – total hardness – pH – iron – calcium – manganese – ORP For some parameters, a design range may be most appropriate.
Operating flow and UVT matrix	Appropriate matrix of paired flow and UVT values based on flow and UVT data.
Operating pressure	The expected operating pressures, including the maximum and minimum operating pressure to be withstood by the lamp sleeves and UV reactor housing.
UV sensors	A germicidal spectral response should be specified. A minimum of one UV sensor should be specified per UV reactor. The actual number should be identical to the UV reactor that was, or will be, validated. The uncertainty of the UV sensors used during **validation** should meet the criteria described in the UVDGM (Chapter 6). The uncertainty of the duty UV sensors during **operation** should meet the criteria described in Chapter 10. Reference UV sensors should be calibrated against a traceable standard. For example, the following standards are currently being used by UV manufacturers: – National Physical Laboratory (NPL) – National Institute of Standards and Technology (NIST) – Deutsche Vereinigung des Gas- und Wasserfaches (DVGW) – Österreichisches Normungsinstitut (ÖNORM)

Table 12-1 Possible content for UV equipment specifications (adapted from USEPA 2006) (*continued*)

Item	Specification Content
Redundancy	The reactor redundancy determined.
Hydraulics	The following hydraulic information should be specified: – maximum system pressure at the UV reactor – maximum allowable head loss through the UV reactor – special surge conditions that may be experienced – hydraulic constraints based on site-specific conditions and validated conditions (e.g., upstream and downstream straight pipe lengths) – maximum flow that maintains sleeve and lamp integrity
Size/location constraints	Any size constraints or restrictions on the location of the UV reactor or control panels (e.g., space constraints with individual filter effluent installation).
Validation	The range of operating conditions (e.g., flow, UVT) that must be included in the validation testing, and submittal of a validation report should be required. The validation testing should be completed in accordance with the procedures and data analysis described in detail in Chapter 6.
UV dose-monitoring strategy	A description of the preferred UV dose-monitoring strategy for the UV reactors.
Operating approach	A description of the intended operating approach for the UV reactors.
Economic and non-economic factors	The necessary information to thoroughly evaluate the UV equipment based on the water utility's specific goals. As appropriate, this information may include both economic (e.g., energy use, chemical use) and non-economic (e.g., future expansion, manufacturer experience) factors.
Lamp sleeves	Lamp sleeves should be annealed to minimize internal stress.

Table 12-1 Possible content for UV equipment specifications (adapted from USEPA 2006) (*continued*)

Item	Specification Content
Safeguards	At a minimum, the following UV reactor alarms should be specified: – lamp or ballast failure – low UV intensity or low validated UV dose (depending on the UV dose-monitoring strategy used) – UV sensor out of calibration – low UVT – high temperature – operating conditions outside of validated range – wiper failure (as applicable) – other alarms, as appropriate
Instrumentation and control	At a minimum, the following signals and indications should be specified: – UV lamp status – UV reactor status – UV intensity – lamp cleaning cycle and history – accumulated run time for individual lamps or banks of lamps – influent flow rate. – UVT measurement At a minimum, the following UV reactor controls (**as applicable**) should be specified: – UV dose setpoints, UV intensity setpoints, or UVT setpoints (depending on the UV dose-monitoring strategy used) – UV lamps on/off – UV reactor on/off control – UV reactor manual/auto control – UV reactor local/remote control – manual lamp power level control – manual lamp cleaning cycle control – automatic lamp cleaning cycle setpoint control

Table 12-1 Possible content for UV equipment specifications (adapted from USEPA 2006) (*continued*)

Item	Specification Content
Performance guarantee	The equipment provided should meet the performance requirements stated in the specification for an identified period or during on-site performance testing. Liquidated damages could be assessed by the utility (if desired) and described in the specification. The following specific performance criteria may be included: – allowable head loss at each design flow rate – estimated power consumption under the design operating conditions – disinfection capacity of each reactor under the design water quality conditions – sensitivity of equipment to variations in voltage or current – UV lamps, reference UV sensor, duty UV sensor, and UVT analyzer (if provided) performance compared to specification
Warranties	A physical equipment guarantee and UV lamp guarantee should be specified. The specific requirements of these guarantees will be at the discretion of the utility and engineer. Significant variation from common commercial standards should be discussed with the manufacturer. Lamps should be warranted to provide the lamp intensity under the design conditions for the fouling and aging factor and a minimum number of operating hours. To limit the UV manufacturer's liability, the guarantee could be prorated after a specified number of operating hours.
UVT analyzer	During operation, the difference between the UVT analyzer measurement and the UVT measured by a calibrated spectrophotometer should be less than or equal to 2% UVT.

REFERENCE

USEPA. 2006. Design Considerations for UV Disinfection Installations. In *Ultraviolet Disinfection Guidance Manual for the Final Long Term 2 Enhanced Surface Water Treatment Rule.*

APPENDIX A

Terms, Units, Symbols, and Definitions

This appendix contains definitions of the terms, units, symbols, and definitions used in this book. Symbols are denoted in round brackets and units in square brackets.

Table A-1 Physical constants of interest in ultraviolet technologies

Constant	Symbol	Value	Units
Speed of light	c	2.99792458×10^{8}	m s^{-1}
Charge on electron	e	$1.602167634 \times 10^{-19}$	C
Planck constant	h	$6.62607015 \times 10^{-34}$	J s
Boltzmann constant	k	1.380649×10^{-23}	J K^{-1}
Avogadro number	N_A	$6.02214076 \times 10^{23}$	mol^{-1}

UNITS AND PHYSICAL CONSTANTS

All of the units used in this book conform to the Système International (SI). An excellent reference for terms and nomenclature in physical chemistry is the International Union of Pure and Applied Chemistry (IUPAC) *Green Book* (Mills et al. 1993). Some of the physical constants of interest are listed in Table A-1.

Other important references are the *Glossary of Terms Used in Photochemistry* (Braslavsky 2007) and Appendix A in Bolton et al. (2015).

The following is an alphabetical list of the terms (with definitions) as used in this book.

TERMS AND DEFINITIONS

absorbance [A or $A(\lambda)$] – decadic logarithm (log10) of the ratio of the incident to the transmitted irradiance as a beam with a narrow set of wavelengths centered on the wavelength λ passes through a medium over a path length l. The absorbance is related to the transmittance $T(\lambda)$ by the expression $A(\lambda) = -\log[T(\lambda)]$.

absorption – physical process of the removal of photons from a beam as it passes through a medium containing absorbing substances

absorption coefficient (decadic) (a) [m^{-1} or cm^{-1}] – the absorbance A divided by the optical path length l.

absorption coefficient (naperian) (α) [m^{-1} or cm^{-1}] – the absorption coefficient (naperian) is used when the decay of irradiance in a medium is expressed as an exponential. Note that $\alpha = \ln(10)a$.

absorption spectrum – a plot of the absorbance [$A(\lambda)$] versus wavelength λ.

action spectrum – the relative UV sensitivity (see germicidal factor) of a microorganism as a function of wavelength.

advanced oxidation technologies (AOTs) – technologies that generate highly reactive intermediates (e.g., hydroxyl radicals) to oxidize and degrade organic contaminants in water.

biodosimetry – the process of validation of a UV reactor by challenging the reactor with drinking water containing a non-pathogenic microorganism.

collimated beam apparatus – a bench-scale apparatus consisting of a UV lamp in an enclosure that allows a narrow beam to be "collimated," either by apertures or a collimating tube. A petri dish containing a suspension of microorganisms is placed under the beam. This is also called *quasi collimated beam apparatus* since the beam is not exactly parallel.

computational fluid dynamics (CFD) – a mathematical model that calculates the paths of microorganisms as they flow through a pipe or UV reactor, taking account of the laws of fluid dynamics.

concentration (c) [M or mol L^{-1}] – the amount (moles) of a substance in solution per liter of solution.

deoxyribonucleic acid (DNA) – a double-stranded helix of nucleotides that carries the genetic information of a cell. It encodes the information for the synthesis of proteins and is able to self-replicate.

einstein – one mole (6.022×10^{23}) of photons.

electromagnetic radiation [J] – energy transmitted at the speed of light (in a vacuum) and characterized by its division into *photons*, which have an energy inversely proportional to the wavelength. Often the abbreviated term *radiation* is used.

energy per einstein (U_λ) [J $einstein^{-1}$] – energy of one einstein of radiation at wavelength λ.

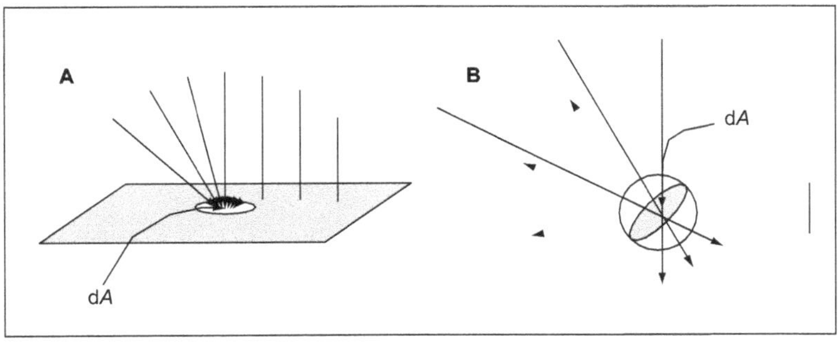

Figure A-1 The concepts of irradiance and fluence rate: irradiance onto a surface (A); fluence rate through an infinitesimally small sphere of cross-sectional area dA (B)

excilamp – a UV lamp that is based on the emission from excited state dimers (excimers) or excite state complexes (exciplexes).

fluence (*F*) – see definition of *UV dose.*

fluence rate (E_o) [W m^{-2} or mW cm^{-2}] (see also *irradiance*) – the total radiant power incident from all directions onto an infinitesimally small sphere of cross-sectional area d*A*, divided by d*A* (see Figure A-1b). Although the unit of W m^{-2} is in use throughout most of the world, the unit mW cm^{-2} is widely used in North America [1 mW cm^{-2} = 10 W m^{-2}].

fluorescence – emission of photons when molecules return to their ground state from excited singlet states of the molecules. The fluorescence occurs at longer wavelengths than that of the exciting radiation.

germicidal factor (GF) – the relative action spectrum of a microorganism normalized to 1.000 at 254 nm. In many cases, the relative spectrum of DNA, normalized to 1.000 at 254 nm, is used as a surrogate for the GF.

germicidal UV radiation – the fraction of the UV spectrum that is effective in inactivating microorganisms. For polychromatic radiation sources, the germicidal UV irradiance is obtained by weighting the absolute irradiance by the germicidal factor.

high-energy radiation – particles (alpha or beta) or electromagnetic radiation (gamma rays) emitted by radioactive substances or high-energy electron beams. High-energy radiation is characterized by causing ionization when absorbed in a substance. This generates high-energy electrons and radicals that cause nonspecific damage to the substance.

inactivation – reduction in the viable cell counts of a microorganism caused by disruption in the ability of the microorganisms to replicate.

irradiance (*E*) [W m^{-2} or mW cm^{-2}] – the total radiant power *incident* from all incoming directions on an infinitesimal element of *surface* of area d*S* containing the point under consideration divided by d*S* (see Figure A-1a). Note that $E = \int E_\lambda d\lambda$, where E_λ is the *spectral irradiance* (unit W m^{-2} nm^{-1}) at wavelength λ. Although the unit of W m^{-2} is in use throughout most of the world, the unit mW cm^{-2} is widely used in North America [1 mW cm^{-2} = 10 W m^{-2}]. The UVDGM (USEPA 2006) uses the symbol *I* for irradiance and calls it *intensity*.

light – electromagnetic radiation in the UV, visible, or IR regions of the spectrum. Note that physicists restrict the definition of "light" to the visible portion of the spectrum.

light-emitting diode (LED) – a solid-state device that emits light on application of a voltage.

low-pressure (LP) lamps – UV lamps containing a very low amount of mercury, such that the mercury pressure in the gas phase is a few Pascals.

low-pressure, high-output (LPHO) lamps – UV lamps with either a large diameter or containing an amalgam of mercury with another element (e.g., gallium). The output of LPHO lamps is 2–3 times that of LP for the same length.

medium-pressure (MP) lamps – UV lamps containing a moderate amount of mercury, such that the mercury pressure in the gas phase is around 100 kPa (1 atm). The output of MP is 20–50 times that of LP lamps for the same length, and the spectral emission is much broader.

molar absorption coefficient ($\varepsilon_i(\lambda)$) [M^{-1} cm^{-1}] – the absorbance of a solution of substance i of concentration 1 M and with a path length of 1 cm at wavelength λ.

nucleotide – nitrogen-containing molecules that link together to form strands of DNA and RNA. DNA contains four nucleotides (adenine, cytosine, guanine, and thymine). RNA also contains these nucleotides, but uracil replaces thymine.

path length (l) [m or cm] – the distance over which a beam of UV radiation passes through a medium.

phage – a virus that infects bacteria (e.g., MS2 colliphage).

phosphorescence – photon emission from long-lived excited triplet states of a molecule as the excited state returns to the ground state.

photochemistry – chemistry induced by the absorption of radiation.

photon – the fundamental *particle* of electromagnetic radiation (e.g., light). A photon has no mass but has an energy inversely proportional to the wavelength.

photoreactivation – the repair of thymine dimer lesions in the DNA chain by the enzyme photolyase activated by near UV and visible light.

radiant emittance (M) [W m^{-2}] (also called *excitance*) of a source – the radiant power emitted in all outward-bound directions from an infinitesimal area dA on the surface of the source. The radiant emittance is a measure of the *brightness* of a source.

radiant energy (Q) [J] – the total amount of radiant energy emitted from a source over a given period of time.

radiant exposure (H) [J m^{-2} or mJ cm^{-2}] – the total radiant energy *incident* from all upward directions *on* an infinitesimal element of *surface* of area dS containing the point under consideration divided by dS (see Figure A-1a). Although the unit of J m^{-2} is in use throughout most of the world, the unit mJ cm^{-2} is widely used in North America [1 mJ cm^{-2} = 10 J m^{-2}].

radiant power (P_Φ) [W] – The *radiant power* of a source is the rate of radiant energy emission or the total radiant power emitted in all directions by a radiation source.

$$P_\Phi = \left(\frac{dQ}{dt}\right) \quad \text{(Eq A-1)}$$

For example, the radiant power of the sun is 3.842 × 10^{26} W. In theory, P_Φ should include all wavelengths emitted by the source; however, P_Φ is usually restricted to the wavelength range of interest for photochemistry. For example, if a radiation source is being used for UV photochemistry, P_Φ would be specified for emission in the 200–400 nm UV range.

radiant power efficiency (η) – defined as

$$\eta = \left(\frac{P_\Phi}{P_E}\right) \quad \text{(Eq A-2)}$$

where P_E is the input electrical power (W) from the wall to run the lamp and its power supply. Sometimes P_E is the electrical power across the lamp.

radiation – see *electromagnetic radiation* and *high-energy radiation.*

RED bias factor (B_{RED}) – a factor to account for the fact that the RED depends on the UV sensitivity of a given microorganism.

reduction equivalent dose (RED) [mJ cm^{-2}] – the UV dose (fluence) delivered by a UV reactor as determined by a biodosimetry test. The acronym REF is also used, meaning *reduction equivalent fluence.*

reduction equivalent fluence (REF) [mJ cm^{-2}] – the fluence (UV dose) delivered by a UV reactor as determined by a biodosimetry test. The acronym RED is also used, meaning *reduction equivalent dose.*

refractive index (for a given medium) (n) – ratio of the speed of light in a vacuum to that in a given medium.

ribonucleic acid (RNA) – an information encoded strand of nucleotides, similar to DNA, but with a slightly different chemical structure.

spectral fluence rate ($E_{o\lambda}$) [W m^{-2} nm^{-1}] – the fluence rate received in a very narrow band of wavelengths centered around a specific wavelength.

spectral irradiance (E_λ) [W m^{-2} nm^{-1}] – the irradiance received in a very narrow band of wavelengths centered around a specific wavelength λ.

transmittance [T or $T(\lambda)$] – ratio of the transmitted irradiance to that of the incident irradiance as a beam at wavelength λ passes through a medium over a path length l. The transmittance is related to the absorbance A by the expression $T = 10^{-A}$.

transmittance spectrum – a plot of the transmittance (T) versus wavelength λ.

uncertainty in validation (U_{Val}) [%] – a factor that takes account of measurement errors made during validation and uncertainties associated with the equipment installed at the utility.

UV dose (fluence) (*F*) [J m^{-2} or mJ cm^{-2}] (also called *fluence*) – the total radiant energy of all wavelengths passing from all directions through an infinitesimally small sphere of cross-sectional area d*A*, divided by d*A* (see Figure A-1b). Although the unit of J m^{-2} is in use throughout most of the world, the unit mJ cm^{-2} is widely used in North America [1 mJ cm^{-2} = 10 J m^{-2}]. The UVDGM (USEPA 2006) uses the symbol *D* for UV dose (fluence).

UV intensity (*I*) [mW cm^{-2}] – this is a term that is used extensively in the UVDGM (USEPA 2006). It generally has the same meaning as *irradiance*, although in some cases it means *fluence rate*. The UVDGM does not define these terms adequately. Note that the unit of W m^{-2} is in use throughout most of the world, but the unit mW cm^{-2} is widely used in North America [1 mJ cm^{-2} = 10 J m^{-2}].

UVA – UV spectral range from 315 to 400 nm.

UVB – UV spectral range from 280 to 315 nm.

UVC – UV spectral range from 200 to 280 nm.

UV-LED – an LED emitting in the UV range.

UVT (%) – percent transmittance of a beam of UV light as it passes through a medium over a path length of 1 cm.

vacuum ultraviolet (VUV) – UV spectral range from 100 to 200 nm.

validation factor (VF) – the factor that must be divided into the RED to obtain the validated UV dose in the UVDGM.

wavelength (λ) [nm] – the distance traveled by a beam of light divided by its frequency.

ACRONYMS AND ABBREVIATIONS

ACGIH – American Conference of Governmental Industrial Hygienists

AOP/AOT – advanced oxidation process/technology

AWWA – American Water Works Association

AwwaRF – Awwa Research Foundation

CF – correction factor

cPEL – ceiling permissible exposure limit

DBP – disinfection byproduct

D/DBPR – Disinfectant and Disinfection Byproducts Rule

DVGW – Deutscher Verein des Gas- und Wasserfaches

GFI – ground fault interrupter

GWR – Ground Water Rule

GWUDI – ground water under direct influence (of surface water)

HAA5 – five haloacetic acids

HSP – high service pumps

I&C – instrumentation and control

IDLH – immediately dangerous to life or health
IESWTR – Interim Enhanced Surface Water Treatment Rule
IUPAC – International Union of Pure and Applied Chemistry
IUVA – International Ultraviolet Association
LP – low-pressure (UV lamps)
LPHO – low-pressure, high-output (UV lamps)
LRAA – locational running annual average
LT1ESWTR – Long Term 1 Enhanced Surface Water Treatment Rule
LT2ESWTR – Long Term 2 Enhanced Surface Water Treatment Rule
MCL – maximum contaminant level
MP – medium-pressure (UV lamps)
NDMA – N-nitrosodimethylamine
NEC – National Electric Code
NIOSH – National Institute for Occupational Safety and Health
NIST – National Institute of Standards and Technology
NSF – NSF International
NWRI – National Water Research Foundation
O&M – operations and maintenance
OCC – online chemical cleaning
OMC – online mechanical cleaning
OMCC – online mechanical chemical cleaning
ÖNORM – Österreichisches Normungsinstitut (Austrian Standards Institute)
ORP – oxidation reduction potential
OSHA – Occupational Health and Safety Administration
PEL – permissible exposure limit
PLC – programmable logic controller
POE – point of entry
POU – point of use
PWS – public water system
RFP – request for proposals
RFQ – request for qualifications
SWTR – Surface Water Treatment Rule
TLV – threshold limit value
TOC – total organic carbon
TTHM – total trihalomethane
UPS – uninterruptible power supply
USEPA – United States Environmental Protection Agency
UVDGM – Ultraviolet Disinfection Guidance Manual
WTP – water treatment plant
WWTP – wastewater treatment plant

REFERENCES

Bolton, J.R.; I. Mayor-Smith; K.G. Linden. 2015. Rethinking the concepts of fluence (UV dose) and fluence rate: The importance of photon-based units – A Systemic Review. *Photochemistry and Photobiology*, 91:1252–1262.

Braslavsky, S.E. 2007. Glossary of Terms Used in Photochemistry. 3rd ed., Pure Appl. Chem. 79(3):292–465.

Mills, I.; T. Cuitas; K. Homann; N. Kallay; K. Kuchitsu. 1993. *IUPAC Green Book: Quantities, Units and Symbols in Physical Chemistry*, 2nd ed., Blackwell Scientific Publications, London, UK, 1993.

USEPA. 2006. *Ultraviolet Disinfection Guidance Manual for the Final Long Term 2 Enhanced Surface Water Treatment Rule.*

INDEX

Note: *f* indicates figure; *n* indicates (foot)note; *t* indicates table.

L

M

N

O

T

U

V

W

Z

www.ingramcontent.com/pod-product-compliance
Ingram Content Group UK Ltd.
Pitfield, Milton Keynes, MK11 3LW, UK
UKHW052012041026
15590UKWH00008B/2284

9 781647 170592